13 Novembre 1989

—

Merci de votre accueil

voyage au Canada.

Je suis persuadé que nous allons faire du bon travail ensemble -

Jag är helt övertygad att vi kommer att bygga upp en mycket gott samarbetet.

Pierre

THE GREAT COPPER MOUNTAIN

THE GREAT COPPER MOUNTAIN

THE STORA STORY

SVEN RYDBERG

Illustrated by Fibben Hald

Published by
STORA KOPPARBERGS BERGSLAGS AB
in collaboration with
GIDLUNDS PUBLISHERS
HEDEMORA
SWEDEN

Graphic design: Christer Jonson
Cover photo: Jan Eriksson, STORA
Paper (bound pages):
Grycksbo StoraFine, 100 grams
Paper (dust jacket):
Grycksbo StoraFine, 135 grams
Flyleaf: Papyrus Colorit
Binding: Corallklot
Printing ink: G-man
Printed in Sweden by Fälths Tryckeri, Värnamo 1988
Bookbinding: Svanbergs, Lessebo
English translation: Jeremy Lamb

Illustrations: Fibben Hald 1988
ISBN 91 7844 113 7

Contents

Introduction

Machine Director Christoffer Klem achieved considerable fame at the Copper Mountain when, at the turn of the seventeenth century, he introduced two epoch-making technical innovations—the series-coupled pumpworks and the water-operated mine hoist. The introduction of these new techniques at the Mine raised its technology to a level fully comparable with the most advanced mining operations on the Continent.

Klem had been summoned to Sweden by King Karl IX himself. The same applied to a number of German copper melters who had been enticed to Sweden at approximately the same time to practice new methods for processing the ore. For Karl IX, the Mine was a prime concern. The fate of the Mountain was of vital interest to the kingdom's ruler.

This was also true when the first major shaft at the edge of the body of ore had reached its estimated lowest point in 1648, at a depth of 120 meters, after being worked for 11 years. The event was considered so important that it was immediately reported to the Chancellory in Stockholm.

Similar examples can be found in the annals of Stora Kopparberg (the Great Copper Mountain) throughout the ages, from the series of royal charters issued during medieval times up to the negotiations about Group structural changes in the 1980s.

Historically, Stora commands a unique position. No other Swedish company is in the same class when it comes to the length of time during which the Company has been considered an operation of national importance.

In a country long characterised by an uncomplicated barter economy, the Copper Mountain early acquired a decisive role in national economics. Only iron and wood have played an equally significant role. The Crown, with its notoriously disastrous finances, was assiduously committed to developing strategies of varying ingenuity, with the objective of squeezing maximum revenues from the Mountain.

Periodically, sections of the ore deposits were pledged to foreign creditors. At other times, all sales were handled by a monopoly. Taxes and other duties were levied with impressive efficiency and considerable imagination.

However, the national significance of the Copper Mountain also had other consequences. Mine operations were subject to the continous and detailed attention of the nation's leaders, who were determined to achieve ever greater efficiency. The nation's foremost experts in mining and melting were engaged to this end. The most successful of the master miners became entrepreneurs on a large scale, and developed a wide range of technical and economic skills. In the farms where they lived, and which dotted the neighboring countryside, people were as likely to debate the nation's trading position and international developments as the day-to-day problems of operating the Mine. Contact between Sweden and other countries had been frequent since the thirteenth century. Production, then as now, concentrated largely on export markets.

The master miners played a key role in writing the nation's history. They also had good reason to keep in close touch with the history being written in other parts of the country and the world.

Mining, and those activities associated with it—raw material supply, ore refining, transport and trading—provided thousands of people with a living, from generation to generation.

Much has changed at Stora Kopparberg over the years. One business activity has grown out of another, revealing a process that may seem almost inevitable when seen in retrospect. Access to ample supplies of timber was, from the very start, a basic prerequisite of copper mining. This also applied to the need for energy. A system of lakes situated in the countryside above the Mine provided a means of establishing the country's first really large-scale water-control scheme. The mining of iron ore developed as a complement to copper mining operations. Despite crises and a number of difficult strategic decisions, Stora's development is perhaps best characterized by the word 'continuity'.

Tradition and renewal both play key roles in this development. It is possible to find clear parallels between the medieval cooperative of Tiskasjöberg and today's worldwide Group.

The national role played by the Copper Mountain has been clearly stressed in the numerous descriptions of the Company published since the 1600s. Nevertheless, this role has normally been marginal to the main narrative. It has been less common to take what is in itself a fairly natural step, and try to place Stora Kopparberg's business operations in a more general perspective related to the political, economic and technical factors which to a large extent determined its development. The following is an attempt, if in somewhat rhapsodic form, to do just that.

During its history, Stora has operated under many different guises: Tiskasjöberg, Kopparberg, Stora Kopparberg and Bergslaget. Over the years, many have simply thought of the Company as 'Berget' (The Mountain).

It is The Mountain which, precisely 700 years ago, was first mentioned in written documents.

In the name of God

TAKE THE ROAD west from Stockholm past Drottningholm, and you come to Svartsjölandet, earlier known as Färingö. Here, in an area overgrown by birch and scrub, you will come upon fairly high stone walls, the remains of a fortress-like building. Once upon a time, this was one of King Magnus Ladulås' many residences. It was here, in 1288, that he had summoned his bishops to a meeting. The meeting comprised Archbishop Magnus Bosson, Bengt Birgersson, Bishop of Linköping, who was the King's younger brother, Anund Jonsson from Strängnäs, Johannes from Åbo and Bo from Växjö, as well as Bishop Peter Elofsson from Västerås.

One of the questions on the agenda concerned a document which was to be presented to Pope Nicholas IV. This particular item involved a matter of principle: there was strong resentment in Sweden about the fact that the Swedish archbishop was junior to his Danish clerical colleague in Lund, and the Swedes were determined to redress the situation.

It had been more than a hundred years since the Swedes, by taking advantage of a favorable opportunity, had succeeded in establishing the right to elect an archbishop of their own. Sweden now felt itself to be on the same footing as more culturally sophisticated Denmark. The matter had become relevant due to the fact that Magnus Bosson had earlier foolishly decided to travel to Lund, where he allowed himself to be

anointed, which at the time had made the King extremely angry.

Bishop Peter presented an item which formally had nothing whatsoever to do with the ruling council, but which was clearly considered to be of the greatest importance.

The background to this case was the fact that he had incurred substantial costs in connection with his installation as bishop in 1284, not least of which would have been the 'annat' (a financial tribute to the Pope), as a result of which he found himself in financial difficulties. At this point, Peter transferred his (the diocese's) interest in the mine at Tiskasjöberg to his nephew, Nicolaus Christineson.

This interest comprised an eighth share in The Mountain. The cathedral chapter had now demanded that this share be reacquired. Nicolaus was to be well compensated. In exchange for his share, he received a farm owned by the Church, which comprised mills, arable and meadowland, pasture, fishing rights and forest land, as well as land in another part of the same parish. Clearly, this was a major transaction. It was only natural that the diocese should become involved in the mining industry: it was quite simply a question of an institution with substantial financial resources in search of a good investment.

The document was intended to confirm that everything had been carried out in an entirely correct manner, and that the transaction had been approved by the diocese. King Magnus himself, as well as the archbishop and his colleagues from Linköping, Strängnäs and Åbo, all approved the transaction by applying their seals to the document, the exception being the Bishop of Växjö, who had not yet received his own seal. At this time, the placing of one's seal at the bottom of a document meant more than mere approval of the contents: it also involved an obligation to ensure that the measures specified in the document were subsequently implemented in full.

Whether everything was really carried out correctly is nevertheless somewhat uncertain. Bishop Peter may in fact have acted rather too quickly. Whatever the cause, subsequent complaints arose that he had high-handedly deprived the diocese of valuable property.

There is little doubt that Bishop Peter was a decisive character. From the fragmentary evidence available, he appears to have been one of the foremost, if not *the* foremost spiritual leader of his time. His exploits on crusades in Finland, carried out during the 1290s, are well documented. Towards the end of his career, he came into characteristic conflict with those in power over the question of ecclesiastical privilege. He died in exile, in Norway, in 1299.

When mention of the mine at Lake Tiska is first made in an official document, it has already developed into a large and nationally important enterprise. It is reasonable to assume that The Mountain was already by this time the largest single workplace in Sweden. It is possible to date the interest of a number of other parties around the time when the deed of transfer was drafted, or immediately afterwards: these included members of the Royal Family, magnates such as Greger Magnusson, president of the courts of appeal of the Västmanland and Dalarna division, as well as other Swedish burghers and merchants from Lübeck. A number of local master miners may also have been involved. The Company had a well-developed organization with a fairly complex ownership structure.

These interrelationships make it clear that mining operations at this time were clearly well established.

When was the first ore mined?

Opinions vary. Well into the eighteenth century, following an early Swedish historical tradition, the start of mining operations at the site was commonly dated far back in prehistory. Perhaps the most picturesque advocate of this tradition is the otherwise relatively unknown bookkeeper Berndt Sveder, who in 1784 claimed that he had documentary evidence which revealed that the mine had first been opened in 1480 BC, and that these early miners had supplied King Solomon with copper to ornament the temple in Jerusalem. Sveder was in good company. Olaus Rudbeck, the most famous Swedish scholar of the seventeenth century, was convinced that Swedish mining traditions dated back more than a thousand years. Even the relatively analytical historian Olof von Dalin believed that Swedish copper production could be traced back to a period long before the birth of Christianity.

But at precisely this point in time—the middle of the eighteenth century—the Danish historian, Jacob Langebek, reached quite different conclusions. Using a combination of analytical skill and convincing documentary evidence, he corrected a number of legendary misconceptions, with the result that several ancient mines, previously thought to have dated from prehistory, found the length of their assumed history sharply reduced. Following Langebek's intelligent and critical studies, the opinion that the mine at Lake Tiska had been opened sometime during the 1280s gained general acceptance. This view was further strengthened by the fact that Bishop Peter's letter indicated that the debated eighth share in The Mountain was acquired by the diocese thanks to his efforts. This fact in itself, does not, however, necessarily imply that he was directly involved when the mine was first opened.

In recent decades, opinions have changed. New methods—spectroscopic analysis of metals, pollen analysis and carbon dating—have provided new insights.

The probable origin of the mine was moved back to sometime around the year 1000 AD.

This was the situation at the end of the 1970s. Since then, the debate has continued. A number of contributions made by the Swedish archaeologist Inga Serning have proved decisive. The situation which she describes may be summarized as follows. In 500 AD, the Sveas (from central Sweden) were a people who not only possessed advanced metal refining skills—they maintained a lively trade with peoples on the far side of the Baltic. Relatively close, in Poland, there is evidence of mining activities which date from between 300 and 400 AD. There is no reason to assume any significant time lag with respect to the dates on which new techniques were applied on the Continent and in central Sweden during the Iron Age. Inga Serning notes that many of the thousands of small mining pits or 'scrapings' found in Sweden's mining districts may well date from the Iron Age. The difficulty of acquiring precise data about these sites is partly due to the fact that a rich deposit was mined on a continuous basis, while mines which failed to yield results were soon deserted, and thus long ago became filled in and overgrown with vegetation. At a burial site in southern Dalarna, dated 600 AD, Inga Serning has also discovered a piece of molten metal which, on analysis, was shown to contain constituents similar to those found in the nearby Grängesberg deposit, constituents quite different from those which characterize the local bog-iron ore.

At the same time, a new analysis of the layers of sediment in Lake Tiska produced some surprizing results.

The assumption on which this analysis was based was the belief that sediment at the bottom of this small, shallow lake would inevitably reflect any pollutant activities which had been located on the banks of its inflowing streams. Polluting effluents would have left their mark in the sediment. The analysis indicated a very sharp change at a level that could be dated to the eighth century. At this point, the metal content of the sediment rose sharply, indicating probable smelting operations—and thereby mining operations—on the mountain above Lake Tiska.

The importance of the results gained from this sedimentary analysis should not be exaggerated. This analysis suffers from a number of weaknesses. Nevertheless, it is worth noting that it has long been known that the inhabitants of the nearby flatlands known as Tuna, and those who settled by the shores of Lake Runn, were involved in relatively large-scale metal processing operations at an early date. They produced more iron than they required for their own needs. Trade with others further south must have blossomed. The island of Helgö, on Lake Mälaren near Stockholm, for example, was an important center for metal crafts and international trade, at least from 500 AD.

It is tempting to conclude that a peasant industry like that found at the southern end of Lake Runn could have been based on the Tiska mine.

The land in these parts, north of Runn, was part of the common land owned by the peasant farmers of Tuna and Torsång.

In this context, it is relevant to note the close relationship between ancient Swedish iron-smelting techniques and those applied in the production of copper.

The basic principles which govern the extraction of copper from a sulfidic ore—and there is no doubt that this is the activity that was being carried out in 1288—are determined by immutable natural laws and thus, in principle, have remained unchanged since prehistoric time. The sulfur is forced out by roasting the ore, once or several times, so that the copper in the ore becomes increasingly concentrated. Normally, it is enough to repeat this process twice. More specifically, the process came to involve the following stages: first, an initial cold-roasting stage, which produced a primary product known as copper matte. This copper matte was then roasted again and then subjected to a final smelting, which resulted in a black end product: crude copper.

In terms of general principles, the process used at the Copper Mountain clearly involved nothing original. However, the construction of the smelter *was* original. It was a low shaft furnace, with breastworks only a meter high and the hearth located beneath the center of the furnace. This type of furnace is found nowhere on the Continent, but the design's lack of a prehearth relates it instead to the traditional Swedish furnace for smelting iron ore. Conversion of this actually quite advanced furnace from iron production to copper smelting has been judged a relatively simple process.

The type of activities suggested could have been carried out by fairly small units and may well have continued over a period of several hundred years, without arousing any special attention.

Nevertheless, it is possible that these activities contributed to the international reputation Sweden acquired as a rich source of metallic ores during medieval times. In his comprehensive encyclopaedia, 'De proprietatibus reum' (The Nature of Things), published in 1240, Bartholomeus Anglicus claims that Sweden was engaged in the large-scale export of silver, copper and iron. Even the rhymed chronicles of the period refer to the country's substantial production of silver. In Olaus Magnus' well known 'History of the Nordic Peoples', written in Latin at the beginning of the sixteenth century, we are told that, if lightning strikes one of the country's ore-bearing mountains, "at the top, on the side or at the foot", it exposes veins of silver which glimmer through the cracks in the rock. "This glimpse," Olaus continues, "attracts people who are consumed by greed, lighting in them unquenchable and steadily increasing flames of desire for possession."

The talk about silver mines has something to do with the Copper Mountain, although the connection is uncertain. There is much to support the view that the earliest mining operations were started in the western part of the ore deposit, where the ground sloped down towards a bog which existed at that time. The foremost historian of the Mine, Sten Lindroth, believes that at that time, the slope consisted of exposed shale, the ore being on the surface in the form of crumbling, easily collected iron-ocher, a reddish type of rock. During the final years of the 1200s, it is likely that mining was carried out in a number of shallow dressing pits dug in the broad pyrite veins found in this part of the deposit.

It has been pointed out that some of the mines in this deposit, when first mentioned in the mid-sixteenth century, were described as silver mines. In the document of 1288, the type of metal to be mined is not clearly mentioned. The possibility that the Copper Mountain was originally opened for the extraction of silver cannot therefore be ruled out. There is no doubt that the ore in those areas which have been pinpointed *does* contain silver.

Even if the deed of transfer of 1288 thus leaves us in doubt over a number of points, it nevertheless provides important details concerning the scale of activities, and the way in which they were organized. The very fact that the deed itself was preserved has, not inaccurately, been described as a stroke of luck.

What *did* get lost, however, were the original "letters patent". However, we do know that one did exist. The letter signed by King Magnus Eriksson, dated 1347, and which is the earliest to have been preserved, is expressly formulated as a confirmation of earlier agreements, made by the King's forebears, which had been lost due to carelessness.

In a collection of letters patent dating from the seventeenth century, the following reference is made to another large medieval mining operation, the Norberg mine: "The first letter to be copied should have been the one granted by King Magnus Ladulås, but since this one is in the Latin tongue it may, in the opinion of this scribe, be passed over."

When Magnus Eriksson describes rights granted by his forebears in the document relating to the Copper Mountain, it seems not unreasonable to assume that he was referring to a document which originated from his grandfather, Magnus Ladulås.

Having come this far, a new question must be raised: Were there any circumstances special to the 1280s which could have made the Mine of sufficient importance to arouse national interest? Or, to put it another way: What was the global situation at this time, in terms of political, cultural and technical developments?

Continental Europe

To help us answer this question, let us revert to the gentlemen at Färingö. When their conversation turned to the state of the world, they probably had numerous viewpoints.

This was the height of the Middle Ages, a highly cosmopolitan epoch. It has been claimed that a federation of European states was never closer to being realized than at this time.

The same culture was shared by all. Throughout Europe, the educated classes enjoyed a common language: Latin. They also shared approximately the same basic principles. This was the Golden Age of the Scholastics, as well as of courtly chivalry. The authority of the Catholic Church was unchallenged. Roman and canonical law went from strength to strength. There was a climate of lively intellectual debate, which bridged national boundaries.

Powerful kingdoms were developing, covering large geographical areas. France, the most powerful country in this part of the world, was one example, England another. Nevertheless, these kingdoms were not 'united' nations in the sense that they were to become later. In France, the Duke of Burgundy maintained his independence from the French Crown, along the nation's eastern border, while the English had settled along the Atlantic coastline. Once again, England was also embroiled in violent internal strife, in an attempt to unite the island of Britain. Wales was conquered, but Scotland was able to maintain its independence in alliance with France. In Germany, Rudolf, the first of the Hapsburgs, became king. He had also succeeded in restoring some degree of order to the country after the chaotic interregnum which followed the fall of the imperial house of Hohenstaufen. Nevertheless, Germany remained a divided country. This also applied to the Iberian peninsula, which was divided into five separate kingdoms, and to Italy. In the north, power lay in the hands of the rich city republics, further south was the Papal State, and in the far south was the kingdom of Naples-Sicily, ruled by a French prince, who had also been crowned King of Jerusalem.

In the East, the power of Byzantium was on the wane. The Russians, who lived under the brutal and arbitrary rule of the Tartars, were generally considered to be outside the European community, and were often lumped together with their Mongolian masters. In the far west, the trading republic of Novgorod enjoyed a special status and was able to maintain a high degree of independence.

Obviously, there were many causes of conflict in Europe and, inevitably, there were wars, if on a fairly limited scale.

The Pope played a unique role in this situation. At the beginning of the twelfth century, the Pope acquired an authority corresponding to that enjoyed by the Holy Roman German Emperor. In the years which followed, his power increased. Pope Gregory VII strongly promoted the traditional concept of the world as a holy state in which the Pope, Christ's representative on Earth, possessed ultimate spiritual and temporal authority. The Pope who proved most successful in promoting these claims was Innocent IV, in the mid-thirteenth century. However, his successors were subjected to increasing scepticism about how much temporal authority should be attached to the papacy. Competition from the rulers of the larger kingdoms became increasingly intense.

Nevertheless, the Church remained a dominant factor in the power game. Its revenues were enormous.

These revenues derived not only from the Church's own property and possessions. Taxes and special contributions were also successfully levied from the world's Christian nations. Kings and princes paid their tributes to the Pope according to the degree to which they accepted his role as a temporal ruler. To these contributions were added donations from individuals who found themselves in difficult situations, especially in connection with the administration of extreme unction. When faced by the fires of Purgatory and the realities of eternal punishment, the only help to be had was from priests who, through their prayers, could intercede on behalf of the penitant. It has been estimated that, in the thirteenth century, the Church owned almost a third of the land in the Catholic countries of the world.

The monastic orders played a key role in the political and cultural influence of the Church. During the 1200s, the older orders of Carthusians and Cistercians were joined by two mendicant orders, the Dominicans and the Franciscans, whose influence spread rapidly.

According to the rules of the Order, the Cistercians were to establish their monasteries in remote areas, and commit themselves to clearing land for cultivation. They have commonly been depicted as unworldly and solitary, tending their herb gardens in a peaceful state of contemplation. Nothing could be further from the truth. These monks were pragmatists, men who knew how to utilize the majority of natural resources available to them, especially those which could generate material benefits. The monastery at Meaux, in France, thus specialized in the breeding of sheep, and by the middle of the thirteenth century had flocks totaling approximately 11,000 animals. The Order had also dominated the production of iron in several European countries for the past 100 years. This was the result of a conscious policy, which originated in France and England but which was subsequently pursued in other parts of Europe in parallel with the establishment of new monasteries. The iron-clad soldiers of the time proved good customers.

The intense interest these monks showed in iron also provides an early and therefore unusual example of how learned men can systematically and consistently become involved in practical technology. In Sweden, too, there is evidence of the Cistercians' interest in processing iron, an interest centered on the monasteries of Alvastra and Nydala.

When at long last, in the 1480s, the province of Dalarna (where the Copper Mountain is situated) established its sole monastery—which was also the most northerly of the Swedish monasteries—it was built by Cistercians at a place which came to be called Gudsberga (God's Mountain). Here as elsewhere, the monks became deeply involved in local ore processing. They acquired most of their ore from the nearby mine at Bispberg.

The Dominicans and Franciscans were known as preaching friars. The main function of both orders was to spread the word of the Catholic faith to the broadest possible public. In Sweden, during the middle ages, the preaching friars' effect on the spiritual life of the nation was so widespread and so deeply embedded that we have difficulty in comprehending it fully today.

The monastic orders maintained lively communication between their hundreds of widespread monasteries and with their controlling organs. This was an important factor in the international distribution of news at that period.

The nobles who gathered at Färingö were probably fairly well informed about the many different aspects of the world around them. The majority of them had travelled to other countries, and acquired personal experience of life on the Continent. Several of them could quite probably look back on many years of study in other countries. They most likely had studied in Paris, which at that time was the unchallenged center of intellectual life in Europe.

Sweden

When it came to Swedish domestic policy, the bishops themselves played key roles. In the Sweden of the 1280s, the Church was, with the King, the most powerful element in the kingdom.

When Magnus Ladulås, after a revolt against his brother Valdemar, allowed himself to be crowned king in 1275, his success was largely due to the fact that he had been able to count on the support of the bishops. His coronation in the following year served as proof that he had received the official blessing of the Church, thus becoming king "by God's grace". The usurper had acquired legitimacy. However, the prelates made sure that they benefited from the situation. In consequence, Magnus made a number of promises in connection with the coronation, which were highly advantageous to the Church. In principle, he waived all royal taxes and levies on the Church's property. All fines imposed in connection with ecclesiastical matters were also to be paid to the Church. This decision established the ecclesiastical nobility.

The distinction between the secular and ecclesiastical nobility in Sweden was fairly flexible. All the bishops during Magnus' reign are said to have been members of the noble families.

At the same time, the position of Sweden's spiritual leaders as a separate group had become increasingly clearly defined as the century progressed. Celibacy was formally made obligatory in connection with the visit made by the Papal Legate, Vilhelm of Sabinas, in the middle of the thirteenth century. But this was only one of the rules contained in the canon law introduced at this time in Sweden.

The Church was involved in the life of the individual from cradle to grave. Baptism, confirmation, confession, the annual celebration of Holy Communion, marriage, extreme unction—these all served to emphasize the individual's dependence on and solidarity with the Church. All important decisions were made in God's name.

During this period, science was unable to offer any alternative to the dogma preached by the priests.

However, the influence wielded by the leaders of the Church was not based entirely on their spiritual authority. They also wielded considerable economic power. The Swedish princes of the Church built more than churches—they also erected palatial buildings, constructed of stone, as personal residences. Equally important, they also surrounded themselves with large bands of armed retainers.

The princes of the Church were superior to feudal knights. A bishop was entitled to an escort of 30 riders. Other members of the Royal Council had to make do with twelve. Only the dukes, all of whom were members of the Royal Family, were allowed retenues larger than those of the most elevated prelates.

King Magnus shared the same ambitions as the Continental monarchs. He also wished to establish a strong, national monarchy. But unlike many Continental princes, he consciously sought the support and blessing of the Church in striving to achieve this goal. This alliance with the Church helped him to counterbalance the powerful families of the worldly aristocracy.

The first years of King Magnus' reign had been filled with strife and intrigues. It was not until some way into the 1280s that he was able to feel reasonably secure in his possession of the crown.

As far as can be judged, Magnus was a highly educated man, well versed in the intellectual trends of his time. His knowledge of Latin is well documented. During the critical initial stage of his reign, the way in which he wielded his royal authority revealed considerable strategic skill, especially with respect to the way in which he charted a safe course between the various nordic power groups.

In all essential respects, he completed the organizational work started by his father, Birger Jarl, in the middle of the thirteenth century.

For a long time, the kingdom had comprised no more than a loose federation of different counties, in which the king represented almost the sole unifying factor. Magnus ensured that the royal castles in the various parts of the country were brought up to standard and manned by strong garrisons under the command of royal bailiffs. National unity was strengthened.

It was during King Magnus reign that medieval society, based on the granting of privileges, achieved its permanent form. Decisive in this context was the statute drafted following a well-attended meeting of magnates in 1280.

The main clause of this statute referred to renewal of laws governing the peace, which had first been drafted by Birger Jarl. The statute also protected farmers from having to provide expensive hospitality to self-invited guests from the nobility, and required each village to provide overnight quarters for travellers passing through. It was also forbidden to demand more taxes from the farmers than the law permitted, or to force them to provide free transport.

The ruling which aroused the most attention was that which exempted the king's retainers and those of his closest supporters from payment of taxes to the Crown, as well as the squires of royal personages and "all those men who serve under the King with their own war-horse".

This ruling has been considered final proof of the establishment of a secular privileged class in Sweden.

The statute derived from the fact that heavily armed mounted armies had by this time become the decisive factor in military conflicts.

The legislative process started by Birger Jarl was also pursued in a number of other areas. The earlier West Göta Code, a code of county laws which originated from the beginning of the century but which was first preserved in a complete handwritten version from the 1280s, was succeeded by a number of other county codes. These included the East Göta Code and the sophisticated Uppland Code, a work to which archdeacon Andreas And contributed, drawing on his knowledge of canonical law, gained while in France. The only city code to have been preserved also dates from this time.

Magnus used to summon meetings, or "diets", similar to that held on Alsnö, at irregular intervals. No definite rule applied as to who should participate, but attendance without an invitation was forbidden.

The group of advisors with which he surrounded himself, and which has already been mentioned, comprising princes of the Church and their worldly counterparts, was a novelty in the Sweden of this period. In addition to a chancellor, Magnus also appointed a Lord Justice and a Lord High Constable as members of the council. At a meeting held in 1284, the council already appears to have become a permanent institution.

At the Skänninge meeting, the laws governing the peace of the realm were extended: a special "royal peace" was proclaimed. This meant that, when the King was visiting a particular county, peace was to be observed "between all men". Anyone who subsequently broke this peace became an outlaw for life.

Unlike the earlier families of the nobility, who were merely of local importance, the nobles during the reign of King Magnus acquired a national status. The most prominent council members owned property in several different parts of the country.

The power of the old noble families was negatively affected. From this time on, they swore an oath to serve as the King's men. The relationship of mutual trust which existed between lord and vassal came to form the basis of the Swedish state.

However, these pledges of fidelity were no guarantee of uninterrupted peace within the realm. Magnus' exercise of his authority twice led to open conflicts, which resulted in the severe punishment of the offenders. Anxiety and a feeling of insecurity about the future of his own line also led Magnus, as early as 1284, to persuade a diet to crown his son, Birger, although only a boy of four at the time. According to an agreement reached with the King of Denmark, Erik Klipping, Birger would subsequently marry his daughter.

Royal foreign policy

The main aspects of Magnus' foreign policy have already become apparent.

In the complex game of politics played by the three Nordic kingdoms, a game in which opposition elements among the noble families of one country (which often had relations and property on both sides of a national border) sometimes sought support from the ruler of one of the neighbouring countries, Magnus consistently sided with Denmark's legitimate ruler, both against opposing elements in Denmark and against Norway.

Like his father before him, he established close cooperation with the "Wend" trading cities in northern Germany, especially Lübeck. When, in 1285, he was elected to act as a mediator between the trading cities of the Wends and the King of Norway, his decision primarily benefited the Germans. In 1288, when he used rather tough means to tighten his control over the island of Gotland and its important trading center Visby, he was supported by Lübeck, and thus also received the support of the cities with which it was affiliated.

For some decades, the Swedes had been trying to gain a secure foothold on the other side of the Baltic. Crusades had been carried out in the Baltic territories, southern Finland and in the country around the Neva estuary. Religion seems unlikely to have been the sole motive for these crusades, which also served commercial and political interests. The contact established with the Russians proved highly profitable.

Communication with more distant European countries was more sporadic. It is perhaps worth noting, however, that the French king, Philip the Fair, presented Magnus with a priceless relic, a thorn from the crown of thorns worn by Christ himself.

The knights

It was during Magnus' reign that the Swedish chivalry blossomed adopting the continental style. A Swedish military class was consolidated, linked by a common ideology. Like so many other aspects of the society of the time, the courtly ideal of chivalry was permeated by the values of the Church.

A true knight was expected, in every situation, to behave as a soldier of Christ. He was to defend the Church, remain fearless in the face of the enemy, generous and magnanimous, merciful to the weak, protect women and never break his vows to his lord. Demands with respect to the more bookish virtues were less severe, however.

It is tempting to ask how the extremely stylized way of life indicated by such a code, so often the subject of laudatory ballads and romances, actually functioned in practice. Many have claimed that this code of behaviour, to a high degree, serves to highlight the yawning gap between ideal and reality, word and action, which characterized this period.

Critics have maintained that the medieval knight was in reality a rough, brutal and ignorant lout, whose talents were strictly limited to certain physical skills. The world in which he actually lived was far more brutal, drab and limited than the ideal presupposed.

Perhaps the situation seemed different to those alive at the time. After all, this was a time when the contrasts between rapture and despair, brutality and compassion, arrogance and humility were much greater than those we experience today.

The King himself

Quite probably, King Magnus himself was, for many people of the time, the Swede who came closest to the chivalric ideal. Judging from the contemporary 'Chronicle of Erik', which provides incomparable and detailed insights into the realm of the time, seen from the aristocracy's viewpoint, Magnus' subjects assessment of his rule was fairly positive, despite a slight qualification: "During all the days of his life he served the cause of righteousness, maintained the peace and chastised with vigor".

Magnus was certainly no paragon of virtue. Compared with other powerful men, however, he appears to have been remarkably devout and fair-minded. Subsequently he was also occasionally referred to as "King Magnus the Holy".

Especially noteworthy is the size of the charitable donations he made to the Church throughout his reign. This applied both to ecclesiastical buildings and to taxation. In particular, this applied to the establishment of monasteries and nunneries. The greyfriars' monastery in Stockholm (the Franciscan Order) received magnificent donations. He donated a large site in the north of the city to the newly-established nunnery of Saint Klara, of which his daughter (still a child at the time) eventually became Abbess. Magnus' will also included a legacy to the grave of Saint Francis of Assisi.

It has been suggested that this consistent liberality hints at the need for some private atonement.

As regards fear for his immortal soul, he was not unlike his father, Birger Jarl, who in his time had, in the form of a specially composed letter, begged forgiveness from the Pope. His brother Valdemar made a special pilgrimage to receive the blessing of the Holy Father in Rome.

Science

The senior members of the priesthood represented more than the nation's faith: they represented learning. The most important position on the ruling council, that of chancellor, was held by a man of the Church, who was supposed to be well-read and blessed with linguistic skills, experienced in the art of preparing documents and formulating decrees. During the 1280s, this post was held by Peter Algotsson. Peter belonged to one of the country's richest and most prominent families, and had many years of study at the University of Paris behind him.

During the final years of the thirteenth century, the most respected academic of this world-famous seat of learning was Thomas Aquinas. This was the man who, beyond all others, had succeeded in codifying the dogma of the Catholic Church, through a fruitful combination of Christian doctrine and the Aristotelian interpretation of the world of nature, based on rationality and experience. The learning of the Scholastics culminated with Thomas Aquinas. His clearly defined and consistent vision of the world remains relevant even today. His teachings still constitute the basic principles of Catholic orthodoxy.

At this point in time, the comprehensive writings of Aristotle had only just become known to the West. The person who perhaps contributed most was Thomas' teacher, Albertus Magnus (who died in 1280), a man who had written copious commentaries on Aristotle. In these, he strongly emphasizes the opinion that Aristotle's authority in matters relating to the natural world preempts that of the Bible and the Fathers of the Church. Albertus' interests ranged beyond mere theology. He carried out in-depth studies in the physical sciences, and was the author of the first European flora (a work describing plant species). Both Thomas and Albertus were known of in Sweden. Petrus de Dacia, a Swedish Dominican friar, had probably attended lectures given by both of them, some of them in Cologne, others in Paris.

The teachings of Thomas Aquinas were not to remain uncontested. One of his opponents was Boetius of Dacia, who may have been a Swede and, if so, was the first Swedish medieval author to achieve an international reputation.

The well-known English academic, Roger Bacon, who was another highly individual thinker, also studied at the University of Paris during the latter part of the thirteenth century. Both as a student of nature and as a philosopher, he is recognized as one of the earliest champions of empiricism. The fact that his writings strongly emphasized the value of practical experience did not endear him to all. He believed that practical experiment offered more than theoretical speculation in the search for truth. Opinions such as those expressed by Albertus and Bacon were uncomfortable and objectionable to the Establishment, even if they were considered stimulating by those of more alert intelligence.

The image of the University of Paris as a peaceful theological college is therefore false, or at least incomplete. Academic curiosity and study involved not only matters of faith, but also the study of nature. Nor were students restricted to the study of pious tracts. Literature of a much more worldly, not to say earthy nature was becoming increasingly available, literature which satirized both the Church and the chivalric ideal with gay abandon.

Although overshadowed by the University of Paris, other educational institutions—such as the medical school at Salerno, the Faculty of Law at Bologna, and the University of Oxford—also played roles of importance. The German-speaking universities were not to flower until the following century.

As for Sweden, as early as the beginning of the thirteenth century, Pope Honorius III had encouraged the country's priesthood to send suitable young men to study in Paris. By the 1280s, there were probably some tens of Swedish students studying in Paris at the same time.

In 1285, Andreas And, already mentioned above, who later became

Dean of Uppsala, bought a house in the Latin Quarter of Paris which was to be used as a place where scholars from Uppsala could meet. However, there was not room for more than twelve at any one time. Some of these scholars were not only allowed to stay in the house: they also received food and pocket money. Others were less fortunate. It was understood that, on their return to Sweden, they would reimburse all costs connected with their studies in Paris, when they had secured a salaried post.

Subsequently, similar boarding colleges were established for students from other bishoprics.

When the student returned to Sweden, he had normally acquired a Master's degree, involving the study of Aristotelian philosophy and logic. Only a small minority of these students would have succeeded in entering the Faculty of Theology. Bishop Brynolf Algotsson, later to become a famous Latin poet, is said to have spent no less than eighteen years in Paris, prior to his return to Sweden in 1277.

Scholarship primarily concentrated on knowledge of spiritual matters. Knowledge of more concrete skills, however, such as medicine, was the subject of strictly limited interest.

Academics were primarily interested in providing theoretical commentaries to the words of the Ancients, such as Hippocrates and Galenos. As for the situation in the Sweden of the time, the Bridgettine monk Peder Månsson, himself a true polymath, was still writing in the following manner as late as the early part of the sixteenth century:

"In Sweden, it is shameful and upsetting to admit that, if anyone becomes sick, this is taken as a signal to get the shovels out to prepare the grave."

He continues:

"Nobody is skilled in the art of healing in this realm." A Swedish surgeon was completely ignorant, "unable to cure even a blemish."

The attempt made by skilled researchers to provide a less simplistic vision of the period has yielded only modest results. It still seems clear that the healers of the thirteenth century had little more than elementary medical books and herbal remedies, containing "lists of prescriptions and good advice for complaints and maladies, ranging from headaches and scabies to infertility."

As in so many other areas, it was primarily the monks who were responsible for any knowledge there was. The Cistercians grew medicinal herbs and introduced the practice of opening veins ("bleeding" the patient), a practice applied as a "cure-all" in cases of internal complaints.

All of this becomes easier to understand if one remembers that at this period it was as natural to rely on the intercession of God or a saint as a means of restoring health as to rely on the healing powers of some medicine.

Technology

The idea that scholarship could be used in the pursuit of secular progress was entirely foreign to the gentlemen of Färingö. The Golden Age lay behind them, not ahead of them.

In the short-term, however, they must have noticed certain improvements. The Europe of the time was characterized by sharp growth in the population and the substantial expansion of trade, based on an increasingly developed industrial infrastructure. A large number of towns were established at this time, assuming one is willing to use such a term to describe what was often little more than an officially recognized peasant village. Nevertheless, the end result was an urban culture which produced a more and more varied and sophisticated range of crafts.

The road system was improved. At the end of the twelfth century, work had been started on surfacing some of the main thoroughfares of Paris with paving stones. A century later, the same thing was done in London.

The network of Roman roads, which had fallen into a state of disrepair, was improved. A few decades into the thirteenth century, it was possible to take pack-horses across the Alps by way of the St. Gotthard pass.

Another way to deal with the problem of bad roads was to shoe the horses with iron, a practice which became increasingly common. Improvements in harness and hame (either of two curved bars holding the harness, attached to the collar of a draught animal) led to dramatic increases in what draught-animals could achieve. It has been calculated that the costs of transport by road declined by two thirds between the time of the Romans and the thirteenth century. A simple but practical development in transport, which also dates from the 1200s, is the wheelbarrow. Once again, as in so many other cases, the Chinese were several centuries ahead of the Europeans.

A revolution was effected in shipping, in that the traditional steering oar was gradually supplanted by the rudder. Unlike the oar, the rudder formed an integral part of the vessel, which enabled the construction of larger vessels and improved navigational accuracy. It became possible to make longer voyages with larger cargoes. In the 200 years between 1200 and 1400, the art of seamanship progressed further than during the preceding 4,000 years.

The most important medieval mining centers on the Continent were located in Saxony, Bohemia and Hungary. The mines in the Harz mountains were well known and were worked before 1000 AD, while the silver mine in Freiburg was worked from the end of the twelfth century. It was during this latter century especially that the Central European silver industry really blossomed, and there was also an increase in the mining of copper.

In Germany, it was common for the lode to be divided up into clearly segregated 'mines', owned jointly by different groups of master miners. The royal prerogative was long established, however: ultimate ownership of the deposits devolved on the Constable of the County. The mineworkers worked with his permission and under his protection.

The actual mining was carried out with primitive implements. The commonest tools were wedges and sledge hammers. The ore was brought up in baskets and buckets. Simple winches were gradually brought into use.

The rich deposits found at Rammelsberg, close to the imperial city of Goslar, yielded silver as well as copper. These deposits provided the basis for Germany's lively domestic and export trade in copper during the 1100s and 1200s, especially that via The Netherlands.

However, Rammelsberg suffered a decline towards the end of the thirteenth century. By this time, most of the surface deposits had been mined, and the lack of efficient pumps excluded the possibility of opening up underground galleries. Operations at other mines on the Continent seem also to have suffered from the limiting effects of this problem, which produced a subsequent decline in the production of copper.

At the same time, technology was being developed which was to revolutionize mining and smelting techniques. The key to (and origin of) this process was the water wheel.

Designs based on a vertical axle came into general use in Europe during the early Middle Ages. They were fairly inefficient, and offered no development potential. On the other hand, large water-wheels with horizontal axles, which became increasingly common towards the end of the twelfth century, *did* have potential. Constructions of this type were already known to the ancients. Initially, they were used for milling grain. In England water-wheels were also used at an early stage to power fulling mills for wool. The installation of water-wheel operated saws proved to be more complex, but had clearly been solved during the first half of the thirteenth century. Evidence which documents the adaptation of the same technique to operate bellows for blast-furnaces dates back to the very beginning of the thirteenth century. It was also approximately at this time that the water-wheel was first used to operate hammer forges.

The new techniques were probably most widely used in the Alps, but they were also adopted in Bohemia, Catalonia, England and France. It was during the thirteenth century that water-wheel operated machinery of various types became widespread in Europe.

The earliest models, powered by 'undershot' water wheels, were increasingly replaced by the considerably more efficient 'overshot' designs.

For the metalworking industry, this development led to a dramatic reduction in the size of the required workforce, while at the same time making it possible to operate on a much larger scale than before.

The triumphant progress of the water wheel in thirteenth-century Europe may be compared with that of the steam engine, following its introduction in the eighteenth century.

The development of the windmill closely reflected that of the water wheel. Both designs could generate between five- and ten horsepower. During this early period, however, the windmill was used only for grinding flour. Windmills were already operating in Sweden by the end of the thirteenth century.

Fundamental to the ability to utilize the new technologies was the know-how necessary to convert a circular movement into a forwards-and-backwards movement, and vice versa. No such technique had been developed in antiquity.

The techniques which now became common, for utilizing power generated by running water and wind, combined with improvements in the harness of draught animals, were enormous advances. If one compares the technology used at the height of the Middle Ages with earlier and more traditional technologies, it is clear that a water wheel or a windmill was able to carry out the work of as many as 100 slaves, the normal power source of ancient times.

Water, wind and physical strength remained the basis of all heavy mechanical work up to the eighteenth century. Similar can be said of the equipment of sailing vessels with rudders. The origins of our modern mechanized society lie in the Middle Ages. The wool industry developed, based on three new inventions: the spinning wheel, the horizontal loom and the fuller's press.

The formula for gunpowder was known to men such as Albertus Magnus and Roger Bacon. The use of gunpowder for guns lay just around the corner. The key to the problem was the ability to produce saltpeter (potassium nitrate) of sufficient purity.

The art of paper-making also spread through Europe. This skill came to the south of France towards the year 1200, slightly later to Italy, and to Germany in the fourteenth century. However, it was still relatively unknown in Europe until the latter period. In the Sweden of the 1280s, paper was still unheard of.

The most spectacular result of medieval technology and craftsmanship are the Gothic cathedrals which appeared in all West European countries, primarily during the latter part of the twelfth century and throughout the thirteenth. All of them involved enterprises which regularly stretched over a period of of several decades, or even centuries, before they were completed. When Notre-Dame, in Paris, was finally completed in the middle of the thirteenth century, no less than ninety years had passed since the first stone had been laid. By the 1280s, work on Westminster Abbey, Köln cathedral and the Church of Maria in Lübeck had already been in progress for a number of decades.

During the latter part of the thirteenth century several large buildings were started in Sweden. Cathedrals in the new Gothic style were built in Linköping, Uppsala and Skara. Construction of Uppsala cathedral started in the 1270s, but it proved to be a time-consuming enterprise. In 1287, a French stonemason and architect, Etienne de Bonneuil, was commissioned to help, and arrived in Uppsala with a troupe of apprentices. The cathedral took more than a hundred years to complete.

The cathedrals were much more than merely concrete expressions of the power of the Church and of the munificence of its parishioners. The proud erection of these soaring Gothic arches is also witness of a highly-advanced technology, a technology which matured under the guidance of the master builders of the thirteenth century. At the same time, builders of the period mastered the craft of making stained-glass windows, designed to enhance the air of solemnity and devotion beneath the high vaults.

Construction equipment was simple and the work was dangerous, especially when heavy items had to be lifted, when everything depended on a primitive lifting system based on fixed tackle and a hoist. Scaffolding consisted of planks lashed together, combined with wooden beams inserted into holes in the walls.

Commerce

Agriculture was also expanding. Generally speaking, it was still the most important means of livelihood. At the height of the Middle Ages, between 90 and 95 percent of the population of Europe supported themselves by working on the land.

A lot of new land was cultivated within the borders of the older West European nations. In addition, there was considerable colonization of Eastern Europe by the Germans, who expanded into countries such as Poland and Hungary.

Something which had a radical effect on Europe, especially Sweden, was German expansion in the north, towards the Baltic coast, where the Duke of Saxony, Henry the Lion, established the city of Lübeck in 1143. It was from here that he started to trade with the Scandinavian countries.

The German colonists continued to expand eastwards along the coast. The weapon they used to conquer the wilderness and its Slavic inhabitants, the 'Wends', consisted primarily of superior technology. Their heavy ploughs enabled them to break up and cultivate the soil. They also introduced a heavier axe and the water wheel. By the mid 1200s, Brandenburg was exporting grain to England.

The Teutonic Order also played a major role in this process of colonization.

The first Orders of Christian knights, the Knights of St. John and the Knights Templar, expanded rapidly and acquired considerable influence. The Teutonic Order was founded in the 1190s. The focus of their operations had gradually moved northwards. By the end of the thirteenth century, the Order virtually held the position of a major power in northern Europe. By the 1280s, it had defeated the Slavic Prussians. It also controlled considerable stretches of land along the Baltic coastline. German feudal lords ruled the country, German burghers filled the towns. In the country, the land was cultivated by German settlers. The indigenous population had either been exterminated or become serfs.

As for trade, the Teutonic Order cooperated with the newly-established towns which had sprung up along the Baltic coastline in the south. Towards the end of the thirteenth century, the Wendish towns, led by Lübeck and in close cooperation with Hamburg, formed what was still a fairly loose federation, a federation which would eventually acquire a more concrete form when it became the Hanseatic League. Riga, with its German traders, and Visby, the majority of whose inhabitants were German, formed important links in the commercial chain, as did the Norwegian town of Bergen.

Together, the federation dominated trade in the Baltic. In the west, its interests stretched as far as London. In the east, as far as Novgorod. The free city of Novgorod played a key role in trading policy, being the only gateway which offered access to Russia and to the roads down to Constantinople. The most important goods traded at this time were furs and hides.

After the turn of the thirteenth century, a new German type of vessel, the 'kogg', was fast becaming the most common merchant vessel in northern Europe. It had a clinker-built, high-sided, broad-beamed and fully-decked hull, about twenty meters long. Single-masted, it carried a square sail and close to 200 tons of cargo. This type of vessel was especially suitable for measurement cargoes such as grain. The construction and charter of a 'kogg' of this type involved a financial commitment beyond the resources of the poor Swedes.

Traffic on the Baltic sealanes became increasingly lively. As a European trading area, it was approaching the Mediterranean in importance.

Swedish trade

For Sweden, the effects of such developments were dramatic. The increasing traffic across the Baltic introduced Swedish products to the North German towns. The cooperation which had been established between Lübeck and Hamburg meant that these products even reached Western European markets, due to the enterprise of German merchants.

Whereas spiritual life in the Sweden of the Folkunga dynasty was dominated by French influence to an extent that was not to be equalled for several hundred years, dependence on the Germans was decisive with respect to economic and social life. The rapid pace of material growth was without question primarily the result of contact with the Germans. But they were not the only ones.

In trade with Russia, carried out via Novgorod, the burghers of Visby—admittedly many of them German immigrants—as well the Gotland sailor-farmers, still played a major role. The Swedes also had direct contact with other countries, such as England.

Swedish exports consisted of skins and fur goods, tar, iron and—at least from time to time—agricultural produce, such as butter. Imports included salt, hops, beer, wine and spices, and cloth.

Calculated as a total figure, overseas trade was certainly of very modest proportions but, generally speaking, the nation became increasingly prosperous. Fixed taxes were introduced, payable in cash, which generated new funds for the treasury. Land rentals were raised. This enabled both the King himself and a number of nobles to amass considerable wealth. In this respect, developments in Sweden paralleled those experienced by many West European nations.

Urban communities started to blossom, not uncommonly comprising a significant number of German merchants. This applied particularly to Stockholm, which took a real leap forward in the 1280s. Magnus Ladulås' encouragement of German immigration was a matter of policy. The Germans responded to this welcoming attitude. The Lübeck historian Detmar calls Magnus "a lover of peace and justice in all its forms". In a letter written by the prior of the Sigtuna dominicans to the Pope in 1289, he describes a Stockholm that in a short time had grown into a well-populated and lively commercial center, to the satisfaction of everyone, including the monks. However, such developments were not unique, applying also to half a dozen other coastal towns.

The inland communities grew in importance in the same manner. Improvements were made to what in many cases was an ancient network of roads. In what is now the southwest of Sweden, roads dating back to ancient times branched off into the interior of the country. One ancient road led from Skara to Jönköping "at the junction of the eight roads". From Jönköping, which was granted formal privileges in 1288, the road continued to Östergötland, where it joined the great trade-route and royal highway which led from Skänninge towards Söderköping. There was also a road which ran from north to south between Örebro and Kalmar. Stockholm depended largely on access by sea, but there were also roads from the Lake Mälaren into the interior of the country. Another important road led along the ridge which stretches north from the region of Västerås towards the Copper Mountain. The Northern Way went from Stockholm towards Gävle and then continued northwards.

The mining-district town of Arboga is first mentioned in 1286. The towns of Köping and Örebro started to grow at the same time—all three responding to the development of the mining industry.

The general purpose of granting a town special privileges was, as on the Continent, to provide support in establishing a special type of community, a center of trade and craftsmanship, distinct from the surrounding rural communities.

The Middle Ages was one of the great periods of agricultural expansion in the history of Sweden. The number of rural communities established at this time remained virtually unchanged until the end of the eighteenth century. The lightweight and simple wooden harrow was still used to break the ground. The heavier iron plough already existed, but was not in general use. Agricultural methods involved the intimate cooperation of the village community.

In addition there were the large estates belonging to those families which had won exemption from taxes to the Crown during the reign of Magnus Ladulås, thus forming the core of the nation's nobility. These estates required large numbers of retainers. One social group which did however decline in numbers was that of the bondsmen or thralls, although it was to take another 50 years, after the 1280s, before thralldom was finally abolished.

Summarizing, it may be said that it was during Magnus' reign that medieval culture achieved its peak in Sweden. No king of his lineage was more successful.

Iron and copper

The last few years of the thirteenth century heralded a breakthrough for the Swedish iron industry. Exportation grew. Naturally enough, most of the iron was sold via the North German trading centers. But there is also evidence of sales to England.

The product traded was known as 'osmund' iron. This name first appears in documents dated 1280. An 'osmund' referred to piece of iron weighing close to 300 grams. These pieces were sold packed in barrels. Osmunds were made by decarburizing pig iron (produced in a blast furnace) on a hearth, to form malleable iron which was then hacked into smaller pieces. This was a product of considerable value.

Archeological excavations made in and around Lapphyttan, not far from the old Norberg mine, have recently shown that blast furnaces already existed in Sweden before 1200. The excavations revealed the remains of the world's oldest known blast furnace. Production capacity has been roughly estimated at between two and three tons a year. In principal, there is no difference between this small blast furnace and the larger German version which was introduced in Sweden in the seventeenth century.

The smelting of ore carried out at sites such as Lapphyttan assumes in turn the existence of a powerful air blast, which could only have been achieved using a water wheel. The documentary evidence indicates that water-powered mills existed in Skåne as early as the 1160s, and in Sweden (Örebro) around the year 1200. The discoveries made at Lapphyttan mean that the introduction of the water wheel in connection with the metal-processing industry must now be dated earlier than has previously been the case. Until these revelations, this technical breakthrough in the mining industry had been dated around the middle of the thirteenth century.

The large number of hearths connected to the blast furnace at Lapphyttan indicate that, even at this early stage, an entire team of mining peasants was involved, who shared joint responsibility for operating the blast furnace, but who then were individually responsible for their production of malleable iron. There seems to be no trace of any German influence in connection with the earliest stages of these mining activities.

What was the situation at the Copper Mountain?

Here, once again, the water wheel must have been introduced for smelting operations during the 1200s. This was a decisive event in the history of the Copper Mountain. Capacity increased, production costs decreased.

The introduction of the new technology, a community characterized by improved organization, increasing prosperity and the accelerating development of new trading contacts opened the way for the Copper Mountain's first Golden Age.

Germans may well have been involved in day-to-day operations. They were certainly involved in marketing the copper.

Increased production meant that, although the small-scale and decentralized operations of peasant farmers remained viable with respect to iron ore, they were taken over by more powerful interests when the more expensive product, copper, was involved, a product which was in addition concentrated to a single large mining operation.

The legislation invoked was, without doubt, the principle of royal prerogative which had been imported from the Continent. This actually applied to many different areas, such as forests and waterfalls, but was to be applied with especial zeal to deposits of precious metals. According to this prerogative, the Crown was considered the ultimate 'owner' of the Lake Tiska Mine. Mining rights could then be granted to a group of master miners, in return for suitable compensation, either a tenth of the profits or a rental fee.

As mentioned above, models already existed in Germany, in places such as Rammelsberg. At the Copper Mountain, however, the Mine remained a single unit. Mine shares were allocated on a proportional basis, all the shareholders being members of a collective, a type of company.

The end product from smelting operations was crude copper. This was largely exported as a semi-finished product to Western Europe. The traffic was controlled almost exclusively by the Lübeck merchants, who had both the capital and the trading contacts. It seems clear that the market was good in view of the decline of Rammelsberg. However, there must have been some competition from Danzig, which marketed Hungarian copper.

A limited amount of the copper was refined and exported to Russia and the Baltic states. These exports seem to have consisted mainly of copper caldrons and the copper sheet used in their manufacture. Purified or 'refined' copper was produced by melting the lumps of crude copper bit by bit on an open hearth, using a powerful air blast, to remove any impurities by oxidation.

The copper was used for a variety of purposes. It was used for articles of everyday use, both secular and ecclesiastical, for tools, as roofing for churches and castles and in the manufacture of weapons. It was used for domestic items such as caldrons, pots and pans, mortars, lanterns, and cups for holy-water. It was also used for figurative work on reliquaries, as well as on rings, buttons, buckles and pilgrims badges. Items such as these were normally made of bronze alloys. This material was also used to crucifixes, censers, candlesticks and candelabra. One especially important application of copper was its use as a raw material for coinage. This even applied to silver coins, which often contained an increasing amount of copper, decreasing the value of the coinage.

The entry of the Copper Mountain into the history books was no mere coincidence. It was the result of series of interrelated circumstances. The development towards large-scale operation was a response to several factors, both Swedish and international.

If so wished, this may be seen as confirmation that the Sweden of this time had become an integrated part of the European community.

Harsh times

SWEDEN HAD LIVED under Gustav Vasa's austere rule for almost twenty years.

The kingdom he inherited had been paralyzed by poverty. The lack of goods had been oppressive, especially the lack of staple products such as salt and hops. The coinage, the disreputable 'klipping', was of inferior quality. Through the uncompromising exercise of power, strict economies, comprehensive control of virtually all commercial activities and great eloquence in defending his own actions, Gustav Vasa was able to build up and subsequently secure his authority. By 1540, he felt reasonably securely established.

One of the projects he then concentrated on was the development of the Copper Mountain.

It was not simply a question of establishing a more efficient system for levying taxes. Both mining and smelting techniques were to be improved. Naturally, this was to be achieved by giving the Crown greater powers.

The King started his own smelting operations, based on the tenth share of the ore supplied by the master miners. These activities were located at the Born homestead, on the west bank of the Falun River. Here, two smelting houses were built, each equipped with two furnaces—a large operation. After only a few years, the Crown also started to mine its own ore. Eventually, this operation became even larger than that run by the master miners.

As well as being in charge of the Crown works, the King's bailiff at the smelting works was also invested with uncontested authority to direct the operations of the master miners, based on the detailed guidelines supplied by Gustav Vasa.

The year 1540 hailed the start of a new era in the annals of the Copper Mountain—that of the Born Works.

This meant that a process of development initiated just over a hundred years previously had now come to an end, after having passed through every stage from one extreme to the other.

At that time—in the 1430s—the master miners had found themselves to be on a direct collision course with the king of their own country. This was the time of the Engelbrekt revolt.

Led by the master miners, the people of the province of Dalarna had suddenly acquired a decisive role in the formation of the nation's destiny. This event was not an isolated incident. During the unrest of the next few decades, the men of Dalarna continued to have a primary role in the political life of the nation.

This situation developed as the result of a conflict of interest typical of the time. An aristocratic party which accepted a monarch who represented the united Nordic nations, with limited authority, ranged against a people's party, with broad support, which hoped for strong national government.

The king who headed the union had his own ambitions, which were not always completely in line with either of these objectives. Erik of Pommerania, the man against whom Engelbrekt led a revolt, was determined to make the Baltic an inland sea which came under his rule. This objective inevitably led to conflicts, especially with Lübeck and the Hanseatic League.

To achieve these objectives and strengthen his authority, the King required reliable men to head the kingdom's regional authorities and a powerful army. For ordinary people, this meant foreign bailiffs and crippling taxation. As a result of this policy, almost all Sweden's counties were controlled by foreigners. The taxes collected were employed outside the country. In addition, the Hanseatic League blockaded the Swedish ports, which not only led to a scarcity of salt but also to an embargo on exports of mining products.

This latter factor was a particularly hard blow to the master miners of the Copper Mountain, and resulted in violent expressions of frustration. This opposition was especially serious, not only because of the considerable financial resources of the master miners themselves, but also because of the substantial support they could raise among the numerous groups of peasant workers in upper Dalarna, who owned little and were therefore highly mobile. These groups of peasants were largely dependent on the work provided by the Mountain for their livelihood. The economic interests of the master miners were also theirs. For much of the time, the men of Dalarna could also count on the burghers of Stockholm as allies, as well as all other groups involved in foreign trade.

The extent of the patriotism shown by the people of Dalarna at this time remains an open question. In any case, it is clear that they were prepared to take great risks to maintain what they considered to be ancient Swedish law and justice. In particular, this involved the question of taxation. At this early stage, however, they had no strong antipathy towards the Nordic union as such.

It may be possible to trace a certain similarity between the conflicts which developed in Sweden and the popular revolts which flared up on the Continent. Nervous Swedish aristocrats drew parallels with the bloody Hussite wars being waged in Central Europe at the time.

The complex political intrigues which were being played out in the decades following 1430 were further confused by the loyalties which sometimes existed between the common people and certain local noble

families and which, in many cases, played a more crucial role than political loyalties based on ideological principles. It was for this reason that, in 1470, the peasant farmers of Uppland found themselves allied with the noble Oxenstierna and Vasa families on the side of the unionists.

In Dalarna, the latter part of the fifteenth century was a time of rising prosperity. It was not only the Copper Mountain which was generating good profits. East Silvberg, a silver mine on the southern border of the plain of Tuna, also experienced a boom.

Several churches in the area were extended and decorated with frescos, often of high quality. The large sanctuary in Stora Tuna, the county center, was completed, and Kopparberg church was converted to a hammer-beam church with a vaulted roof and brick columns.

A Guild was established at the Mountain, based on the model current on the Continent and in southern Sweden, with Saint George as its patron saint. It differed from the merchants' associations in adopting an independent and militaristic profile. For the members, loyalty to the Guild superceded the law of the land, and the word of the Master of the Guild weighed heavier than that of a judge.

As for political conflicts, the men of Dalarna adopted a relatively consistent stance, even though they sometimes also found themselves on the side of the unionists. Theirs gradually became a much more radical national movement. A number of successful armed confrontations, especially the triumph at the battle of Brunkeberg (near Stockholm) in 1471, raised their self esteem.

Characteristic of the times were the personal loyalties which, from the mid 1460s, united the common people with the members of two leading noble families, both named Sture. For a long time, Västerås Castle was the home of a member of the Sture family, by reason of which he also became Governor of Dalarna County.

The oldest of the line was Nils Sture. He was followed by the three Stures who, in their role as regents, governed the country, apart from a brief interim period, from 1471 until 1520. All of them were to a large extent relying on the country people of Dalarna. For support from abroad, they looked towards Lübeck.

Under the last Sture (1512-1520), this dependence on the people was especially marked. Like many continental regents of the time, he consistently sought support from the people as a means to balance the power of the Council. His regime has even been described as a people's dictatorship.

The people of Dalarna's influence peaked between 1520 and 1521, during a period of Danish ascendancy, when Kristian, the union king, triumphed. The leader of the national opposition, Gustav Vasa, was on the run, and naturally sought refuge in Dalarna, from where he was able to launch the final stage of the liberation of the country at the head of a peasant army. This time it was more than a question of defending local interests. This time, it involved no less than a permanent change in the status of the entire kingdom.

When Gustav Vasa came to power, it was hardly surprising that the people of Dalarna felt they had acquired special status compared to the rest of the country's inhabitants, a status which gave them the right to adopt an especially familiar manner in dealings with the King whom they had helped to the throne. This self confidence developed into something akin to arrogance. Three uprisings erupted in close succession (1524-1531). The main reason for dissatisfaction was a scarcity of goods, the poor coinage and the high rate of taxation, which the King blamed on the need to repay large debts to Lübeck. In addition to this, the new Lutheran teachings were unpopular. The inhabitants of Dalarna listened willingly to those who opposed the king, men who linked their cause to the traditions of the old Sture Party.

The threatening letter issued in 1525 by the people's county council, which gathered in Tuna, became famous in this connection. By way of introduction, the King is reminded (somewhat callously) of how he had once wandered around in the forests of Dalarna, an outlaw—"like a squirrel in the pines" (to quote the poet Karlfeldt)—without a single friend among the Danes or the Germans. He is then reminded how the poor men of Dalarna and other country people hurried to his rescue, and placed him as the "sun in the firmament of the realm". The charter then expands eloquently on what were felt to be the King's errors and his broken promises to the people. The document concludes with a threat: "We respectfully submit this document to You, in all good faith, that You may know that, unless You are able to remedy these wrongs as soon as possible, reject traitors and love the true men of Sweden and...reduce the price of grain...that we can no longer give You the loyalty we have promised."

Approximately at the same time that the Bell uprising (which was the last) erupted, the King found himself fully occupied with the troubles which arose in the wake of the deposed King Kristian II's attempt at reinstatement. But once this conflict had been resolved, it was time for a showdown with the men of Dalarna.

Representatives of the people were summoned to a meeting at the Copper Mountain in February, 1533. The king arrived at the head of a strong military escort. Surrounded by soldiers, the people who had gathered were forced to listen to a ringing indictment delivered by the king and his council. The king was especially enraged about the demand, raised by the country people on the advice of the leading master miner, Måns Nilsson, that the king should ask to be granted safe conduct before entering the county. In future, the king announced, Dalarna would either become an obedient part of the country—or a devastated part of the country.

Despite the amnesty which had earlier been granted, some of the suspected leaders of the uprising were summarily executed, while others were removed to Stockholm, where they eventually met the same fate.

Those sentenced were among the richest men of the Copper Mountain, and had considerable local influence as employers of large groups of peasant laborers from the northern part of the county. Several of them had supported the king during earlier crises. This was true of a number of

the most notable, men like Måns Nilsson, Anders Persson and Ingel Hansson.

It has been maintained that the Dalarna uprisings were fairly mild affairs, more the expression of a wish to put pressure on the king than anything more serious. It never came to more than a bit of sabre rattling. The Bell Uprising acquired special distinction primarily as a result of its detailed description by Peder Swart, the king's personal chronicler, a description clearly written as propaganda. The savage sentences meted out give the impression of an almost personal act of revenge. The king wished, once and for all, to silence the 'rumblings' of the self-important men of the Copper Mountain.

The method proved effective. The role of the people of Dalarna as a political force seemed at an end.

When the Crown estate at Born was established seven years after the king's penal expedition, the pacification process was complete. When the king revisited the Mountain in 1542, he was greeted by an audience of highly attentive subjects.

Sweden, 1540

The degree of Gustav Vasa's authority is also apparent in a number of other areas.

In 1540, Sweden was formally proclaimed an hereditary kingdom. This was the first time in the nation's history.

The ecclesiastical reformation was completed with the implementation of the new Church statutes, by which the nation's priests became state officials. In future, opposition to the king was equated with opposition to God himself.

The fragile relationship with Denmark, which at the start of 1540 seemed likely to lead to open war, improved considerably during the year. An agreement about a pact was drafted and later ratified by both kings.

The vital relationship with Lübeck also developed favorably. In his efforts to gain independence from Denmark, Lübeck had been Gustav Vasa's sole ally. But the price for this support had been high. In addition to the cash debt of 120,000 Lübeck marks which was to be paid, he had been forced to agree to grant a whole series of privileges which gave the merchants of Lübeck an almost total monopoly on Swedish foreign trade. Following a feud in the early 1530s, the city had lost its traditional position of power. The role of economic guardian which the city had enjoyed up to this time, so humiliating to Sweden, could now be dissolved. At a well-attended meeting of the Hanseatic League, held in 1540, the participants agreed to use peaceful means in seeking an agreement with Sweden.

The standard of living in the realm at this time was relatively satisfactory. The diet was undeniably monotonous, comprising mainly grain, meat, fish and butter. Much of the food was old, dry and heavily salted, and had to be swilled down with gigantic quantities of beer. Nevertheless, it provided the inhabitants with an average calory intake which was almost as high as that enjoyed by Swedes at the beginning of the twentieth century.

Sweden was at this stage still very much a peasant economy. No more than 5 percent of the total population of about 700,000 people lived in the towns. Approximately the same number were employed by the mining and metalworking industry.

In terms of foreign policy, Sweden in 1540—and for that matter throughout almost the entire reign of Gustav Vasa—remained an isolated realm. Political activities were restricted to neighboring countries, comprising those nations which bordered on the Baltic, and The Netherlands. Cultural life was also provincial—and remarkably meager.

The international scene

In Europe, the opposite was true: the times were characterized by bubbling vitality, both with respect to science and the arts. In Rome, the aging Michelangelo was in the process of completing his colossal fresco of the Last Judgement on the domed ceiling of the Sistine Chapel. In Venice, Titian was at the height of his fame. North of the Alps, Hans Holbein held a similar position.

Vesalius, a professor at the University of Padua, published his first anatomical charts in 1539. They were revolutionary, since they were based on the dissection of human bodies. The descriptions given of the body's functions and organs by the doctors of antiquity, which had remained unchallenged for so long, now had to be reviewed.

At the same time, the authority of the Ancients was roundly criticized by Paracelsus, a versatile natural scientist, who was the first to practice medicine by administering chemical substances. He attempted to cure syphilis, a newcomer to Europe with a preparation based on mercury.

Nicolaus Copernicus also belongs to the first half of the sixteenth century. As early as 1530, he had already completed the manuscript for his great work on the movement of the heavenly bodies. This work claimed that the Sun—not the Earth—was the center of the universe. Just as the studies of Vesalius had provided a new and different insight into the functions of the human body, Copernicus presented his contemporaries with a new view of the universe. Two crucial pillars which supported the Ancients' view of the world were about to collapse.

On yet another plane, Vanoccio Biringuccio had produced the earliest detailed manual on mining and metalworking techniques. It was published under the title 'Pirotechnica' in 1540, a year after the death of its author.

The Swedes of the early 1540s were probably almost completely ignorant about all these developments.

However, they were likely better informed about political developments, such as the dramatic expansion of the Spanish and Portugese empires.

As early as 1494, with the approval of the Pope, Spain and Portugal reached an agreement as to how to divide the newly discovered parts of the world between them. The border was to stretch from north to south on a line which lay 3,700 kilometers west of the Cape Verde Islands.

Around 1540, the Spaniards were fully engaged in establishing themselves in their various spheres of interest. Mexico, Central America and Peru had already been conquered. Now it was Chile's turn.

The direction of Portugese interest was for the time being focused primarily towards the East. The Portugese established a network of fortified trading posts in Africa, India and even further afield, in the Malacca Straits.

Around 1540, the Spaniards' new wealth, in the form of gold and silver bullion, had as yet had only a limited effect on the European economy, but rumors about this wealth must have spread far and wide. The import of exotic goods by the Portugese, especially spices, played a far more significant role than is immediately apparent from the perspective of a modern observer. The people of the time were great lovers of strongly spiced food. This was of considerable importance in countries where people had to use old or rotten raw materials, as in Sweden. Spices were in fact so important, that they could actually lead to war.

The Swedes of the 1540s were probably fairly well informed about the interrelationships of the various powers on the Continent of Europe.

The endless conflicts which arose between the Hapsburg Emperor, Charles V, and the French king, Francis I, also had a certain effect on the Nordic countries.

Charles ruled an empire which extended over a greater geographical area than that of any previous European ruler. It comprised The Netherlands (including today's Belgium), the land of his youth, a land whose two languages he spoke, which were (and remained) his native tongues. It also comprised Spain and all its colonies, the Holy Roman German empire, the traditional hereditary lands of the Hapsburgs to the southeast and, finally, parts of northern Italy. As a result of the geographical distribution of this empire, France felt threatened from most points of the compass. However, Charles' empire was highly heterogeneous, and difficult to hold together. The factor which unified the countries of this widespread empire was—or at least, according to the emperor, ought to be—the joint belief in the Roman Catholic religion. In this respect, he was to become deeply disillusioned: the evangelical reform movements were to be the cause of endless conflict.

This was also the time of the Ottoman Empire's most ambitious period of expansion in the West. In 1529, Sultan Suliman was at the gates of Vienna. In 1540, the house of Hapsburg lost Hungary to the invading Turks. France signed a treaty with the Turks, and the Emperor signed one with the Persians. The time was now past when the forces of Christendom stood united against the threat posed by 'The Unbelievers'.

The Nordic monarchs had special reason to keep in close touch with the actions and statements of Charles V. The legitimate ruler of the three Nordic countries, Kristian II, was brother-in-law to the Emperor and had his support—albeit halfhearted—in his attempt to win back his crowns. The situation became acute when, in 1531, Kristian landed in Norway with a small force. The result of this minor invasion was, as has already been mentioned, disappointing, but it created a feeling of insecurity which remained, and it was clear that both Sweden and Denmark had reason to see France, the enemy of Charles' empire, as an ally.

The Reformation

There is little doubt that Sweden was an interested spectator of developments on the Continent with respect to the work of religious reformation. Ten years earlier, in 1530, the religious division in Germany had become a fact. It was at this time that the evangelical creed was formulated at a meeting in Augsburg, a creed which was later to form the basis of Lutheran dogma. At approximately the same time, the evangelist German dukes grouped themselves under the Schmalkaldenian Federation for their mutual protection. The teachings of the Protestant reformists spread rapidly.

It was inevitable that the Roman Catholic Church should take countermeasures. It was clear that the astonishing series of corrupt, depraved and cynical power politicians which had characterized the See of St. Peter must be broken. Aspects such as the commercialization of indulgences and the mumbling of masses and requiems in return for payment had to be stopped.

It was the election of Pope Paul III in 1534, in himself a not particularly virtuous example, which was to herald the start of the Counter-Reformation. In the year 1540, Pope Paul established the Jesuit Order. The Inquisition, which was already active in Spain and Portugal, developed new and terrifying aspects.

It was typical of the spirit of the times that even literary works concentrated on religious matters or Church politics.

In this area too, as in the the work of religious reformation, Luther was a central figure. His translation of the Bible, completed in 1534, has been described as an epoch-making achievement. This period, more than any other, was the Age of Bible Translations.

The new humanism developed in parallel with the reasoning of the evangelists. Humanism involved an essentially heathen view of the world. Its core was the teaching of a morality based on the ancient philosophers. But supporters of this persuasion during the early part of the sixteenth century also took to criticising the Catholic Church. One man who was to become especially prominent in the humanist movement was Erasmus of Rotterdam, whose witty and scathing satires of the papacy were widely distributed. They even amused the Pope, who was not particularly zealous himself.

Erasmus died in 1536. By this time a bridge had been built joining humanism's world of ideas, inspired by the ancient philosophers, and the Reformation's belief in sin and salvation. This had been achieved almost solely through the efforts of one man, Philipp Melanchthon. As a professor at Wittenberg, he molded the University into a scholarly focal point for the entire Protestant world.

Sweden had several highly-placed churchmen who possessed considerable knowledge of developments on the Continent. During the years around the turn of the sixteenth century, especially large numbers of Swedes had visited Rome.

One alert commentator of events in the papal city for a long time was the versatile Peder Månsson, a priest and man of science, who lived there as superintendent of the House of the Holy Birgitta between 1508 and 1524. After his return to Sweden he became bishop of Västerås. It was his investment of Laurentius Petri as archbishop in 1531 which secured the apostolic succession in Sweden. Politically, Peder was a nationalist, but he was also a firm adherent of the Roman Catholic faith. He died in 1534.

The Bishop of Linköping, Hans Brask, had also had experience outside Sweden, and had gained a doctorate from a German university. He was one of Gustav Vasa's most trusted assistants during the first few years of his reign, although he later came into conflict with the king.

The two brothers, Olaus and Johannes Magnus, are typical representatives of the international churchman of the Middle Ages.

A number of younger priests had been deeply affected by the teachings of Luther and Melanchthon, many of them, like the famous Olaus Petri, first coming into contact with their ideas in Wittenberg.

The Reformation in Sweden

For Gustav Vasa, commitment to an evangelical orientation was from the beginning determined by the opportunity it represented to improve the national economy by the acquisition of Church property. It gradually became an increasingly ruthlessly used instrument of power.

In the Sweden of the 1520s, there was no general support for the introduction of the Reformation. The new teachings were forced on an unwilling populace. It was therefore important that those who wished to free the Church from Rome avoid confrontation as far as possible. It was a question of taking it step by step. The monastic orders were required to pay additional taxes, but were not yet banned. Prelates sympathetic to the type of reforms suggested by Erasmus were appointed as leaders of the Church in Sweden.

The spokesmen for these new and increasingly clearly Protestant ways of thinking were archdeacon Laurentius Andreae, who at an early stage was appointed the king's chancellor, and Olaus Petri, who in 1524 was invested by the king as secretary to the municipality of Stockholm, and official preacher at Stockholm's Storkyrka (Great Church).

Supported by these two men, Gustav Vasa was able to persuade a meeting of parliament, held in Västerås in 1527, to approve a program which meant, among other things, that 'superfluous' property owned by the Church and the monasteries should revert to the Crown and that "nothing but the unsullied words of God" should be preached. The phrasing was such that these declarations could in principle be accepted by reformist Catholics. However, although this was not a question of enforced conversion to Lutheranism, it did represent a decisive step towards a break with Rome. This result was not achieved without a fight. The king was forced to bring matters to a head by threatening to abdicate if parliament refused to grant his wishes.

Over the next few years, things developed apace. By 1540, the break with Roman Catholicism was complete.

Swedish culture

It was in the same year, 1540, that the king's old retainers, Larentius Andreae and Olaus Petri, were sentenced to death. As it turned out, the sentences were commuted to pardons, but in future it was others who were responsible for the political initiatives taken with respect to the ecclesiastical life of Sweden.

This reaction may be seen as an indication of Gustav Vasa's well-known paranoia. Both were frank and outspoken men, who never hesitated to express their opinions, even if these involved criticism of the king and his actions. Olaus had, for instance, preached a sermon "against dreadful oaths", a sermon which the king, not without reason, felt was directed at him, and by which he felt insulted. But the charges of treason would have been difficult to justify.

The judgement may also be taken as an illustration of the conditions to which spiritual life was subject in the Sweden of Gustaf Vasa.

The country's cultural stratum had always been thin. During the Middle Ages, only a limited number of learned men were to be found in the nation's cathedral chapters and monasteries. For them, the Reformation spelt the loss of the majority of their material resources, while at the same time the ancient inherited learning they honored was condemned as superstition. The monastic libraries were dispersed and destroyed. The churches' missals and bibles met the same fate. Medieval manuscripts were used as wadding for canons or were stapled together to be used as ledger covers. The nation disinherited itself.

The country's bishops complained that the schools had "throughout the realm fallen far behind". They stated that, where you could previously find as many as 300 pupils, there were now barely 50.

Uppsala University had ceased to function even before Gustav Vasa came to power. During the first few years of the new regime, the number of Swedish students at foreign universities also declined catastrophically.

Throughout the entire sixteenth century, with the exception of the year 1525, the country had no more than one printer active at the same time. The total number of books published in Sweden during the 100 years of the sixteenth century corresponded to less than one year's production from the presses of Paris.

As an individual, Gustav Vasa had no interest in education in any profound sense of the word. However, he was in need of competent administrators, advisers with foreign languages and reliable scribes in his chancellery. When in 1540 the king's widely-respected adviser, Georg Norman, who was Pomeranian, drafted a proposal for new guidelines on education, the importance of a thorough grounding in Latin was stressed.

As for studies at university level, Gustav Vasa thought it was cheaper to send students to study at foreign universities than to reestablish Uppsala University. He would simply have to live with the shame of ruling a kingdom without a single higher seat of learning.

In Sweden as in many other countries, the Bible was translated into the native tongue. During 1540, this great task was nearing completion. It was published in the following year.

The Gustav Vasa Bible was Sweden's foremost intellectual achievement of the century. Linguistically, it was a masterpiece. The later editions, both the Bible published under the reign of Gustaf Adolf and that of Charles XII, meant only limited revisions. The original translation remained in use, almost unaltered, until 1917. The greatest honor for this work of biblical scholarship must go to Archbishop Laurentius Petri.

Apart from the Archbishop himself, his brother Olaus was the most prominent author of the period. Initially, his work consisted of clearly phrased and eloquent statements in support of the evangelical faith as taught by Luther. Olaus Petri was also an important historian. His 'Chronicle of Sweden' was primarily intended as a pedagogical work. History was supposed to provide valuable lessons.

However, the way in which it was presented was both factual and relatively critical. During the sixteenth century, no other historian was to attain the same standard of scholarship. The Chronicle, the first part of which was completed in 1540, contributed to his dramatic fall from grace in the same year. The king was interested in the production of useful propaganda: he did not want statements which represented the author's personal opinions or beliefs.

Olaus was ready with his famous 'Guidelines for Judges' by the beginning of the 1540s. They formed the introduction to the Swedish statute-book until well into this century.

This was also the time at which the exiled Swede, Olaus Magnus, published his pioneer work, the Carta Marina, in Venice. Cartography blossomed in the wake of the great geographical discoveries of the period, but this was the first time that the out-of-the-way Nordic countries had been described in a reasonably satisfactory manner, with numerous—although fairly often inaccurate—details.

Swedish politics

It has already been noted that Sweden during the reign of Gustav Vasa was an isolated kingdom. There were a number of reasons for this.

One of these was the king's personal lack of knowledge of how things functioned in other countries. His total experience of foreign countries was limited to his short stays in Denmark and Lübeck, as prisoner and then fugitive, during his youth.

One may speculate about his view of the world around him, based as it was on second- and third-hand information received from his advisers. They brought him little intimate knowledge of the ways of other countries. This inexperience must have been one of the key factors which resulted in such a passive foreign policy. In addition, there were only a very limited number of Swedes who could be trusted with a mission of international diplomacy. Gustav Vasa was almost without exception forced to rely on Germans, whose qualifications for the job differed widely.

The smallness and fragility of the organization is demonstrated by the situation which developed in 1535, when the German secretary to the king fled the country. The king had no option but to rely on his own shaky German when dealing with foreign correspondence.

Another limiting factor was the way in which Gustav Vasa had gained power. Seen in constitutional terms, his election as king was the result of downright rebellion against a legally elected monarch. As far as many Europeans were concerned, Gustav was and remained a usurper. His claim to the throne was never generally accepted. Throughout his reign, he had to contend with his image as a peasant lad who had used violence to get his hands on a crown. Seen in this context, his endless claims about how, fired by selfless zeal, he had faced great and terrible dangers to rescue the nation from ruin are easily understood.

It was also a question of Gustav's personality. He was—it has been claimed—"always easily angered, found it difficult to forgive, and was highly sensitive to criticism, as well as almost abnormally suspicious of everyone and everything" (Wieselgren).

In conclusion, it may be said that Sweden's extremely vulnerable strategic position hardly encouraged the idea of more ambitious initiatives.

With respect to domestic policy, Gustav Vasa sought as best he knew how to copy the autocratic methods of princes on the Continent, assisted by his German advisers.

The letter which the king addressed to the country people clearly shows how things were at this time. Gustav compares himself to Moses, leading the Children of Israel. He warns his subjects of the consequences of their ingratitude to him, and brands those who complain about taxation. It was, he stressed, the king's exclusive right, in the name of God, to give commands and establish laws for his subjects, both with respect to the government and religion.

Finance and trade

Gustav's authority was not built solely on the administrative and military powers which were now available to the Crown. It was also based on the fact that the general situation in the country had stabilized and that the people, as has already been hinted at, enjoyed a fairly high standard of living.

Despite incessant wars and rancorous doctrinal disagreements, the sixteenth century was a period during which living standards rose in many parts of Europe. The houses built in the towns became larger, more imposing and more comfortable, and sometimes even had glass windows and tiled ovens.

Clothing became more elaborate, and the quality of food improved, even for the poor. Artists and craftsmen found themselves in great demand.

Manners also became more sophisticated. The art of eating with a spoon spread rapidly throughout Europe. The fork, however, was still used mostly as a utensil for serving food.

Economic policies were characterized by the European princes' wish to promote their own industries through the imposition of taxes and new regulations. Their aim was also to facilitate exports and provide urban communities with greater opportunity for development. Ultimately, what it really involved was supplying the prince with greater revenues to support his armed forces and his court.

The focal point of trading activity moved from the Mediterranean to Western Europe. The Netherlands became the most densely populated country in this part of the world. It was to here that wine and salt were exported from Spain and France, where they were exchanged for grain, which probably came mainly from the lands of the Teutonic Order, but also for hides, furs, iron, tar and timber from the North.

The Netherlands was also the most significant industrial center in Europe. The most important industrial activity was the manufacture and trading of textiles. The country around Liége was the foremost coal district of the period, linked to a growing metals industry, both in and around the town.

England exported pewter and lead, but the most remarkable growth was noted by the textile industry, based on widespread sheepfarming. The quality, however, could not compare with that achieved in The Netherlands.

In Central Europe, the mining industry during the 1530s was rapidly nearing a peak. This applied to the silver mines in the Tyrol and to the Hungarian copper mines. The art of working and forming metals developed.

In this, as in so many other areas, the large trading houses played a key role. The foremost of these was owned by the Fugger family, from Augsburg, whose fortunes reached a highpoint in the decades up to the mid-sixteenth century. Apart from industrialists, they were also the foremost bankers in Europe. They financed coronations and wars, the Emperor being their most important client. With effect from 1524, the family was granted the right to mint its own coinage.

At this time, Germany was the largest iron producer, with an annual production in the region of 30,000 tons at the beginning of the century. Annual production in France was around 10,000 tons. The market was growing, especially due to the wars being waged by armies which consumed ever increasing quantities of iron.

For long distances, transport by sea was the dominant choice. Shipbuilding was a vitally important craft.

In the 1530s, the Venetians' large trading galleys which, despite their name, were actually sailing vessels, were the most prestigious in Europe. They could carry a cargo of 250 tons and sailed routes from the easternmost Mediterranean to England and Bruges.

However, the galleys met increasingly stiff competition from the fully-rigged and far more maneuverable vessels which were known as 'carracks'. The long ocean voyages of the time were commonly made by ships of this type. There were also a number of other types of trading vessel, some of them of more than 600 tons.

During the first half of the sixteenth century, the English acquired a special status among shipbuilding nations during the reign of Henry VIII, by building a powerful fleet of large ships. The largest of these displaced as much as 1,500 tons and carried as many as 200 canon, with a crew of 900 men. Building large ships became a question of prestige. From the beginning of the 1530s, Sweden also possessed the resources to build ships of this size.

The craft of the shipbuilder was veiled in secrecy. No manuals existed. It was important for the designer to be able to copy as well as he could models which had already proved successful. The key to the problem—and the starting point—was the midships section. The rest was more or less sculpted, according to the builder's skills and knowledge.

Despite the long voyages undertaken by these vessels, navigational methods remained fairly unsophisticated. Medieval sailors relied on compasses, astrolabes and tables of the sun's declination, combined with some simple maps. Techniques existed to determine latitude, but not longitude.

Technology

There is evidence which suggests that, in the first 50 years subsequent to the invention of the printing press in the middle of the fifteenth century, a total of no less than twenty million copies of books were published. This figure was more than all the scribes had succeeded in producing during the preceding 1,000 years. The majority of these works were published in Germany and Italy.

Schoolbooks comprised the greater part, although a significant number were practical manuals covering subjects such as agriculture, gardening and various handicrafts. A substantial number of these manuals focused on mining and metal pocessing.

There was rapid development in industries of this type during the period, which was in particular a result of increased mechanization. This in its turn required greater capital investment than had previously been the case. It was therefore quite natural that trading houses like Fugger's were to play key roles in this development.

Pirotechnica, a work of Vanoccio Biringuccio, already mentioned above, provides a good description of the latest metals technology of the 1530s. It covers mining and mine buildings, a series of smelting processes for different ores, metals and alloys. But the author does not stop here: he also describes how bellows should be constructed and how to assay metals, as well as techniques for minting coins, casting canon and making glass.

Despite his contribution, Biringuccio's treatise was soon to be overshadowed by another work, the *De Re Metallica* of Georgius Agricola, the result of twenty years work which had started in 1533. This masterly description of the many different aspects of metal processing surpassed everything previously written on the subject, and was to be the main work of reference on mining and metallurgy over the next 200 years.

The *Art of the Miner*, by Peder Månsson, who has already been named above, represents the oldest complete description of the copper smelting process in existence. It was also a work of international standing. It was written at the beginning of the century, and although never printed, several handwritten copies were distributed round Sweden.

The first half of the sixteenth century was a time when complete suits of armor were considered the height of fashion. The knight on his charger was supposed to be clad in iron from head to toe. Ideally, his horse should be protected in much the same way. There was a natural interplay between the suits of armor worn and the type of firearms used. In the long run, this equation created an impossible situation for the rider, who found himself having to don more than 45 pounds of additional equipment. The full set of armor was best reserved for parades.

The great breakthrough in hand-held firearms came at the turn of the sixteenth century. In principle, virtually all the main types of rifle bolt to be used over the next 300 years had been developed by the year 1520.

In the 1530s, a new, heavier and more powerful firearm, the musket, made its entry on the scene. A musket charge was in theory powerful enough to penetrate a suit of armor, but since it was a smooth-barreled weapon, it was highly inaccurate. To hit the enemy, it was literally necessary to "see the whites of his eyes" before firing. But it was not long before rifled barrels came into use.

Sweden was among the frontrunners in this development. The troops of the younger Stures' were already equipped with firearms, although cold steel and crossbows were more common.

Canons are first mentioned during the early part of the fourteenth century. In Sweden, they were already in use at the time of the Engelbrekt Uprisings. They had now passed the experimental stage. Charles V made an effort to standardize the weapon with respect to caliber, shot weight and length of barrel.

Iron, bronze and brass were all used for making canons. Although iron was cheapest, and was suitable for casting heavy pieces of artillery, the material also had a number of disadvantages: iron canon could rust, and could develop dangerous fractures. This was one of the reasons that bronze canon were considered superior, although they became worn relatively quickly through the effect of the ammunition.

Each canon was cast separately. This involved great skill. The barrel was seldom perfect. As a result, each canon was apt to possess special characteristics. It was therefore advantageous if the manufacturer could also operate the canon on the battlefield.

Developments in the design of artillery inevitably affected the way in which fortresses were constructed. It was no longer primarily a question of erecting thick defensive walls. It was more a question of designing fortifications which offered the defenders an optimum field of fire. Fortresses were complemented with fortified firing positions, ramparts and moats.

Sweden's trade and economy

In Swedish history, Gustav Vasa is often named as the first representative of the New Age. It seems unlikely that he saw himself in these terms. His trading policies were actually more consistent with the late Middle Ages.

One especially noteworthy aspect of Gustav Vasa's trade policy was his "covetous attitude to goods". Earlier experiences had taught him the importance of 'stocking up', to secure the nation's basic supplies.
Although Sweden was largely self-sufficient, a number of key items still had to be imported. Most important, as already noted, were salt for the preservation of meat and hops to make beer, and cloth.

Imports were encouraged. Exports were strictly controlled. The king felt that foreign trade ought to be left to foreign merchants, instead of going to all the trouble of travelling to other countries oneself, with all the associated risks. He believed that Sweden could command a better price for its goods if foreigners had to travel there to sell and buy. The king was at the same time worried that simple-minded and inexperienced Swedish traders would fall easy victims to cunning foreigners, such as merchants from Lübeck and Denmark.

In the 1530s, this 'covetousness' was expressed in terms of bans on large numbers of exports. This applied to oxen, for instance. They were not to be driven south to the provinces of Denmark, but north, to Sweden's mining district. Export bans were also placed on the largest breeds of horse, which the king needed for his armed forces, and on silver, which the king wished to amass in his own treasury.

Gustav's view of urban development derived from an old domestic tradition, yet he was also open to ways of thinking popular in other countries. The prototype on which the Swedish model was based derived from the relatively independent, rich and favored cities of the Continent. However, what was actually possible, given the rustic realities of Sweden, was necessarily on a much more modest scale. From Gustav's point of view, the network of regulations established in connection with this urban development offered the additional benefit of an opportunity for increased control over a substantial part of the national economy.

Other aspects of the national economy were also affected by a medieval attitude. Gustav Vasa was the son of a large landowner or, more accurately, farmer. He understood how to organize a multifaceted domestic economy. He now ran the whole of Sweden in the same way.

Mining and metal processing

Thus, in many respects, the king was clearly both provincial and conservative. But when it came to mining and metal processing, his attitudes were different. In these two areas he worked hard to achieve improvements and to introduce new methods.

Swedish iron was popular and occupied a secure position as one of the nation's most important exports, despite the fact that it was occasionally superceded in terms of sheer value by other products, such as butter. Around 1540, approximately 3,000 tons of 'osmund' iron was sold to foreign buyers through Stockholm. The largest purchaser was Lübeck, which accounted for more than a third, closely followed by Danzig. The other importers were the merchant cities of the Eastern Baltic, foremost among these being Königsberg, Riga and Reval.

There is no evidence as to the sort of quantities exported from other ports, but what is clear is that the Stockholm region dominated this trade.

Mainly in Danzig, but also in a number of other places, 'osmund' iron was processed into bar iron, after which the refined product was sold to customers in Western Europe. The thought of starting domestic production of bar iron was obviously tempting to the Swedes. Gustav Vasa first personally referred to bar iron in 1539. The trend was towards increased production of bar iron, although the pace at which this occurred was hardly rapid.

Mining and smelting were carried out in small, widespread units. The mines that were worked were all located in central Sweden. Many of them had been exploited for years. Since there were so many deposits in the mining district, while the quantities required were limited, mining regularly took the form of shallow pits. Miners chose to start again at a new site rather than take all the risks involved in mining at depth. The miners were seldom heard to complain about a lack of ore.

However, the supply of timber, which was equally important, sometimes proved a more serious problem. The most common cause was the ruthless exploitation of easily accessible forests.

Mining was a seasonal occupation practised by groups of peasant farmers. Normally, it was limited to the summer months.

The ancient technique of direct smelting in a hole in the ground or on a slope continued, although to an increasingly limited extent.

The more advanced blast furnaces were normally semi-submerged, built into the bank of a small river which powered the bellows. They were usually constructed on a simple timber foundation, rising to a height of about five meters. The amount of power required was minimal. A miner's furnace could be located on the bank of virtually every stream in areas where wood or ore was available.

The forges were also small. As a rule, they had only one hearth and a stretching hammer. Gustav Vasa advised his smiths against trying to build more pretentious constructions. If they were built large enough for two stretching hammers, they would look as if they "were some sort of cathedral". This might prove far too expensive for the ordinary peasant smith.

The iron which the miners smelted from pig iron on their small hearths, and which was then broken into lumps of 'osmund' iron, was often of high quality and low carbon content, with a uniform structure which thus ensured good formability. However, the process involved considerable burnoff, and was therefore relatively expensive.

As far as can be ascertained, all iron manufactures required by Sweden were produced within the country's own borders. The iron industry's own tools, such as hammer heads, anvils, pokers and shovels, were all made in Sweden. This also applied to wire and chains, and especially to military equipment, including gun barrels, iron shot for canon and musket, iron sheet for suits of armor, and sword-blades.

Sala Mountain

Gustav's pride and joy was the silver deposit at Sala. It was a relatively new mine. Large-scale operations had first started in 1510. Hope then ran high, and people flocked to Sala in large numbers with a view to exploiting its riches. In 1540, the mine was just at the start of its short boom, a period during which it achieved an annual production of approximately 515,000 marks of the metal. The silver extracted from this mine exceeded the value of the nation's total iron exports, and was many times the total yield from the Copper Mountain. The timing of this boom was also favorable, since it occurred before the really large quantities of Spanish silver had started to depress the European market. And finally, the country gained considerable prestige from having its own silver production.

The mining techniques applied led to the creation of large open chambers. In the sixteenth century, this way of working, known as the 'pingenbau' technique, was already being replaced on the Continent by a more

modern technique based on shafts, adits and deans (gallery heads). One of the obvious disadvantages of the earlier method was that supporting columns and abutments of ore-bearing rock had to be left behind. All too often they were made too thin, weakening them and creating the risk of serious cave-ins.

Hand-operated winches were used for hoisting, although it is likely that around 1540, they started to construct horse-drawn hoists at the edge of the mine.

Sala posed special problems due to its flat landscape and limited water resources. As late as the 1530s, the bellows at Sala Mountain had to be operated by foot.

On the whole, smelting operations were based on the Continental model. German specialists were summoned to help. They were the best in Europe. The operation of the mine was divided between the Royal Bailiff and individual master miners, although all the ore was to be sold to the Crown.

The Copper Mountain

Thus, the Sala Mine was the subject of rejoicing during the reign of Gustav Vasa, operations at the Copper Mountain were not. The conditions which applied in this case were entirely different. Both the organizational structure and the way of working had been determined long before.

By the mid-1300s, control of mine operations had already been transferred from the aristocratic cartel of the high Middle Ages to a group of local master miners. There must have been many reasons for this, one of which may have been the difficulty of controlling such a many-faceted business from a distance. Another reason may have been that the preconditions for coordinated large-scale smelting operations did not exist at the time.

In the fifteenth century, it was the royal bailiff who supervised mine operations at the Mountain. He was supported by a council, comprising fourteen master miners, who shared responsibility for operations with the bailiff. Two members of this council were elected as Mine Masters, whose role at this time was virtually that of judge.

All the mines and mine chambers were to be made available to the master miners in proportion to the number of shares they held in the Mine. They were permitted access to the various mine faces according to a timetable based on the annual 'gamble'.

The work in the Mine was subsequently based on a daily schedule which was to remain in force for centuries. At the end of the day, firewood was thrown down into the Mine, accompanied by loud shouts of warning, and then stacked against the various faces and ignited. It was important to ensure that all these fires were lit at the same time, after the miners had evacuated the area. The remains of the fires were cleared away the following morning, after which the miners could get to work on the face itself, which had been made brittle by the heat, using wedges and sledge hammers. Once this stage was completed, the chunks of rock had to be cleared aside, before the next team started to throw their firewood into the Mine.

The ore was hoisted by the same simple means as used at Sala.

After roasting, each master miner smelted the ore in his own wholly- or part-owned smelter. These were small constructions, which were operated at least in the spring and fall. The single furnace was kept going by two bellows, which functioned alternately. The leather required for the bellows was the most expensive item in a furnace. Six entire ox hides were needed to construct two pairs of good-sized bellows. The leather's useful life was about two years. On replacement, the old bellows-leather was reused to make gloves, shoes or harness for horses.

It appears that the oldest smelters were sited in the old agricultural district near Lake Runn. They were grouped together in rows by each stretch of water. It was only gradually that people started to build smelters closer to the Mountain, along the stream which ran down towards the Falun River and, from the start of the sixteenth century, on the banks of the Falun River itself. It was important to find a place with a suitably strong current: the distance for transport to the smelter was less important. Smelting techniques during the Middle Ages appear to have been relatively efficient. At least, no major or radical changes were introduced until the beginning of the seventeenth century.

It is clear that the forests close to these operations were already being ruthlessly exploited as early as the end of the fifteenth century. The master miners were forced to purchase timber and firewood "from the farmers".

The greater part of the labor force at the Mountain must have been hired. There is also evidence to suggest that this involved fairly large numbers of laborers, an assumption supported by the anxiety expressed over the possibility that this workforce might "decide to band together". Apparently it was at times difficult to recruit laborers, since asylum was granted to outlaws who wished to work at the Mountain. Given such circumstances, the difference between working conditions at the Mountain and forced labor was probably not that great.

In all, it is probable that several hundred laborers worked at the Mountain. By the end of the fifteenth century, the Mine had approximately 100 master miners.

It has been suggested that annual production exceeded 200 tons of crude copper from the mid-1400s until 1490. Towards the end of the century this figure rose, and there is good reason to believe that annual production increased by some hundred tons. This estimate is based on import figures from Lübeck. Whatever the precise figure, it is clear that the Mine at this time yielded very satisfactory results. This may have been why the regent, Sten Sture, decided to take personal control over the county of Västerås, including Dalarna, in 1493.

This prosperity did not last, however. Only a few years into the sixteenth century, the Mine entered a period punctuated by a number of serious cave-ins. The master miners worked hard to clear the deposits in

order to continue operations, but the Mine had suffered a "mortal wound". Around the year 1510, production had fallen to no more than a few hundred kilos.

One theory which attempts to explain this chain of events suggests that the record production noted in the 1490s was the result of rapid and ruthless mining at depth, which created a number of very large chambers, a careless and exploitative approach which was ultimately paid for by the subsequent cave-ins. Drainage of the Mine had also certainly proved to be a problem. Before the start of mining operations, a large part of the deposit had previously been covered by a marsh.

Seen from a contemporary viewpoint, the mining operations carried out during the latter part of the fifteenth century involved a considerable production volume. It was quite adequate to explain the fairly large number of wealthy gentlemen who were actively engaged in operations at the Mountain at this time.

The situation in 1540

Seen in terms of the amount of metal extracted, the results of King Gustav's energetic efforts to improve operations at the Copper Mountain during his lifetime seem modest. On the other hand, he succeeded in introducing radical organizational changes, which included much greater working discipline and more clearly defined tax regulations. Adopting a highly personal approach, Gustav ensured that the Mine was operated according to his own ideas, and that the results were reported in detail.

This provided the first clear historical account of conditions at the Copper Mountain. In fact, according to Sten Lindroth, the mines at Falun and Sala are probably unique in that they are the only ones of this period whose fate can be traced in such detail. With respect to the Copper Mountain, this refers to a period of 80 years. Of particular interest are the reports submitted by the royal bailiff of the Born estate.

Both the bailiff himself, Nils Larsson, and his 'henchman', Karl Månsson, who were appointed to manage the industrial estate, were first and foremost administrators. This was also true of Nils Geting, who is first mentioned in 1540 with the title of Mine Bailiff. None of them had any particular technical qualifications.

The 'pair' (the traditional name for the smelter furnace with its pair of bellows) was the standard unit used in assessing the size of the master miners' shares in the Mine. The 'pair' could be broken down into smaller units, the most common of which was the 'fourth part', which actually referred to the fourth part of a quarter, or a sixteenth of a 'pair'. In the 1540s, the Mine was operated by approximately 20 'pairs'. The total number tended to vary, as did the number of master miners. In 1540, it seems likely that there were about 80 master miners at the Mine. That same year the king commanded that the smelting operations of the master miners, which at that time involved about 30 separate furnaces, should be discontinued and replaced with fewer but larger installations which were to be sited closer to the Mine. This would enable continuous operation in the same manner as had already been adopted for the Crown's smelters on the Born estate, and would increase yield. This decree did not lead to anything, either in 1540 or later, when the king raised the matter again. It has been suggested that the reason for this lack of action was the inherent conservatism of the master miners. The most decisive reason was perhaps even simpler than this, and lay in the fact that the proposed reorganization was impracticable, due to the lack of adequate supplies of water.

At this time, the Mine was being worked in two large open casts. One was the the original Great Mine (Blankstöten), which yielded an ore consisting of almost pure iron pyrite with an insignificant amount of copper pyrite. The other was the younger Farmer's Pit (Bondestöten), which yielded a harder, copper-rich ore. Both were now mines of considerable size. The Great Pit was a gaping hole, while the Farmer's Pit was worked via a series of 'levels', which lay one on top of the other. The deepest points of both mines were close to 100 meters below ground level.

In addition to these, a new hard-ore deposit was opened up around 1540, called the Johannis Mine. To this was added a number of shallow open-cast 'pits', which were worked in and around the main deposit.

A document dated 1542, which refers to the production of 100 tons as a decline compared with the preceding year, provides an indication of the Mine's approximate production level at this time.

The mining techniques used at the Copper Mountain were basically the same as those adopted at the Sala mine. Instead of a maze of passages, as on the Continent, the Copper Mountain contained great chambers which spread beneath curving, vaulted ceilings, an underground cathedral made possible by the hardness of the Mountain's rock.

The reason Continental methods were rejected lay in the nature of the body of ore itself. It was a great chunk of pyrite ore, quite unlike the veins of ore which had to be followed on the Continent.

King Gustav was always anxious to recruit German mining expertise for Sweden—and with good reason. His aim was to gain knowledge of improved technology. Primarily, he was interested in horse-drawn winches, which were relatively complex constructions, comprising a main beam, cable basket and disc. When these constructions were adopted at the Mountain in the 1540s, it is probable that Sala was able to provide some insights about their operation. These winches, which were used for hoisting of both rock and water in buckets were a major step forward.

Other new inventions, such as treadwheel and pumps, were not introduced at the Copper Mountain until some decades later.

The copper trade

One special reason for Gustav Vasa's decision to acquire such firm control of the Copper Mountain around 1540 may have been the prevailing situation with respect to the international copper trade.

Some of the copper exported by the Fugger family, from deposits in

Central Europe, was transported by a southern route via Venice, while the rest was sent on a northern route from Danzig via the strait of Öresund to The Netherlands. It was around 1540 that the Fugger contract with the Hungarian mining industry ran out. There was some doubt as to whether it would be renewed. A condition for renewal was that the Fuggers, who were bankers to the Emperor, would be granted free passage through the strait between Denmark and Sweden. But as has already been mentioned, the Emperor supported his brother-in-law, Kristian II, and was therefore in conflict with Kristian III, who had been king of Denmark since 1534. Both Denmark and Sweden were at this time in the process of moving towards an alliance with Francis I of France.

The Fuggers hesitated. Their close relationship with the Emperor created obvious difficulties when it came to trading via the Baltic. In such a situation, it might be useful to establish contact with the opposition, in this case the Swedish copper industry. There is perhaps some basis for thinking that Gustav, seeing the possibilities offered by such a situation, determined to uprate the production facilities in Falun.

However this may be, it is clear that the Swedish Crown did not wish to restrict its export trade to Lübeck and the other Wendish and Baltic ports. The Crown was also interested in increasing its number of direct contacts with the West, especially with The Netherlands.

Exports normally passed through Stockholm, but around 1540 a significant amount was being exported from Gävle, on Sweden's Baltic coast. The reason for this was the key role played by Gävle in supplying the Mountain with various types of goods. The master miners paid in copper. At this time, production from private shareholdings was considerably greater than that of the Crown, perhaps seven times as much.

A not inconsiderable part of the Mountain's production was sold by the master miners to individual buyers in Sweden. Typical of these were the coppersmiths. Russian pedlars formed another special category of customers. Limited amounts of copper were also purchased by merchants for subsequent sale to customers in Norway.

It has previously been noted that the quantity of copper mined in 1540, and later during Gustav's reign, was extremely limited. It is difficult to visualize the tremendous efforts that must have been required by the master miners and their mineworkers to achieve even such modest results. They had learned to face up to the constant threat of cave-in and flood over many years. Now they had to contend with the fact that the ore was extremely poor, containing a mere one percent of copper. This meant that the volume of rock which had to be mined and processed, and the quantities of firewood and timber that had to be transported to the Mine were much greater than had earlier been the case. It was not until the 1560s that the miners finally struck a richer deposit. Production crawled slowly upwards, to reach just over 150 tons a year.

The king's energetic and concentrated efforts to improve operations at the Mine paid off in the long run. When a particularly rich body of ore was struck around the middle of the 1570s, combined with the fact that mine operations had become much safer, the organizational structure and technical skills were already in place, providing good conditions for dramatic growth.

1650

A fragile greatness

"ON NEW YEAR'S EVE, the master miners stood in front of the Scale and weighed copper until after midnight." More than 40 weighers were at work. Production for the year totaled more than 3,000 tons of raw copper.

The year was 1650. The Mine Master reported that the Mine had been greatly blessed by the Lord and that the poor master miners were delighted. Hoists and winches worked night and day, but despite this they were barely able to raise all the available ore to the surface. Resources were stretched to the limit. More than ten chambers were being worked at the same time. The Mine Master was unable to open up any more.

Since the turn of the seventeenth century, when production had been about 450 tons a year, production had risen in an extremely encouraging fashion with each decade. Nevertheless, production in 1650 was abnormally high. Annual production averaged over the century was closer to half the 3,000 tons reported for this particular year.

The ore extracted since the start of the century had been comparatively rich—about 4 percent raw copper. But the position had been even more advantageous during the 15-20 years towards the end of the sixteenth century. Suddenly, both the Crown works and the master miners had been able to manufacture more than double the previous level. The ore extracted at this time produced 7 percent raw copper.

But the good result was also essentially a consequence of improved technology. Following the pattern prevailing throughout the Continent, the water-wheel-driven 'pipe-and-ball' pump for drainage was introduced at the beginning of the 1570s.In the 1590s, the Mine had acquired in Christofer Klem a mine engineer who fully mastered the technology of the time. His pumps and hoists laid the basis for the coming greatness of the Mine.

In the 1620s, the simple walking beam was introduced to transfer power to the various pumps. Towards the middle of the seventeenth century, the Mine had already in most respects become, even by European standards, a remarkable and impressive place of work. Thomas Christersson's map of the Mine dating from 1650 provides a good picture of the situation; the two major work sites were still the two vast open casts (Blankstöten and Bondestöten) with smaller finds further east, both shafts and scrapings, either in operation or collapsed. Around the openings to the mines and shafts were hoists, winches and water-powered pumps.

Above the Mine there was a reservoir from which water used in operations was channelled in wooden pipes to the wheel houses.In the middle of the area was a high bell-tower.

The classic eye-witness reports of the Mine at the height of its greatness come from Charles Ogier, a member of a French embassy to Sweden in the mid-1630s. In his diary he writes: "We were truly amazed on our arrival at the Mine opening. What a sight spread out before us! An abyss of incredible breadth and depth opened up in front of us, and around it a row of hand rails so that no-one would accidently approach the edge of the pit and, in terror at such great depth, swoon and fall in. Even when supporting yourself against these rails, everything goes black and you become dizzy as you turn your gaze downwards; but when you get used to it you see the people coming and going at the bottom of the Mine, they look like birds or, more correctly, ants, so small do they appear. Wherever you turn you see wonderful things, both in themselves and in comparison with each other: fire and ice, light and darkness—everything interfused. One would think that it was the chaos of old, such a diverse and confused impression is engendered by the whole scene.

"Our eyes and senses were fully occupied in revelling at the sight, when we suddenly saw one of the workers throw himself down the rope by which the ore, using wheels and hoists, is winched up from the depths of the Mine: not without a shudder of apprehension did we see him so fearlessly let himself slide down". The group went down into the Mine: "Here we came to a big open area, and went from there under an arch formed of ice and vitriol crystals. The ice was thawing, partly by itself, and partly due to the heat from sporadic fires which the mine workers had lit. This caused us to walk under a shower of droplets. It is otherwise both a wonderful and terrible sight, these caves where the flames from the fires which thaw and erode at the rock masses, gleam and flash on the walls. In other places there are a multitude of wheels and machines by which the masses of water from the bottom of the Mine are transported, water which makes it difficult for the workers and which would be inexhaustible if it was not continuously pumped up".

Ogier has given yet another report of his visit, found recently, that to a certain extent supplements the information contained in the diary. Here, he finishes off his description of the Mine itself in the following way: "For those of you who wish to create a picture of the Mine in its entirety, imagine a dark hole, terrible and deep, down to 60 or 70 fathoms (a little over 110 meters), dug out and arched artificially and in different directions, held up by nothing other than itself, filled with fires in different places, filled with smoke and sulfur and the smell of metal, filled with dripping water. And then, in the depths of the earth, black people, like small devils, echoes from hammers and crowbars with which the stone is broken. The cries from the mine workers, from those transporting the ore to the baskets and finally the desolation and the thunderous roar which

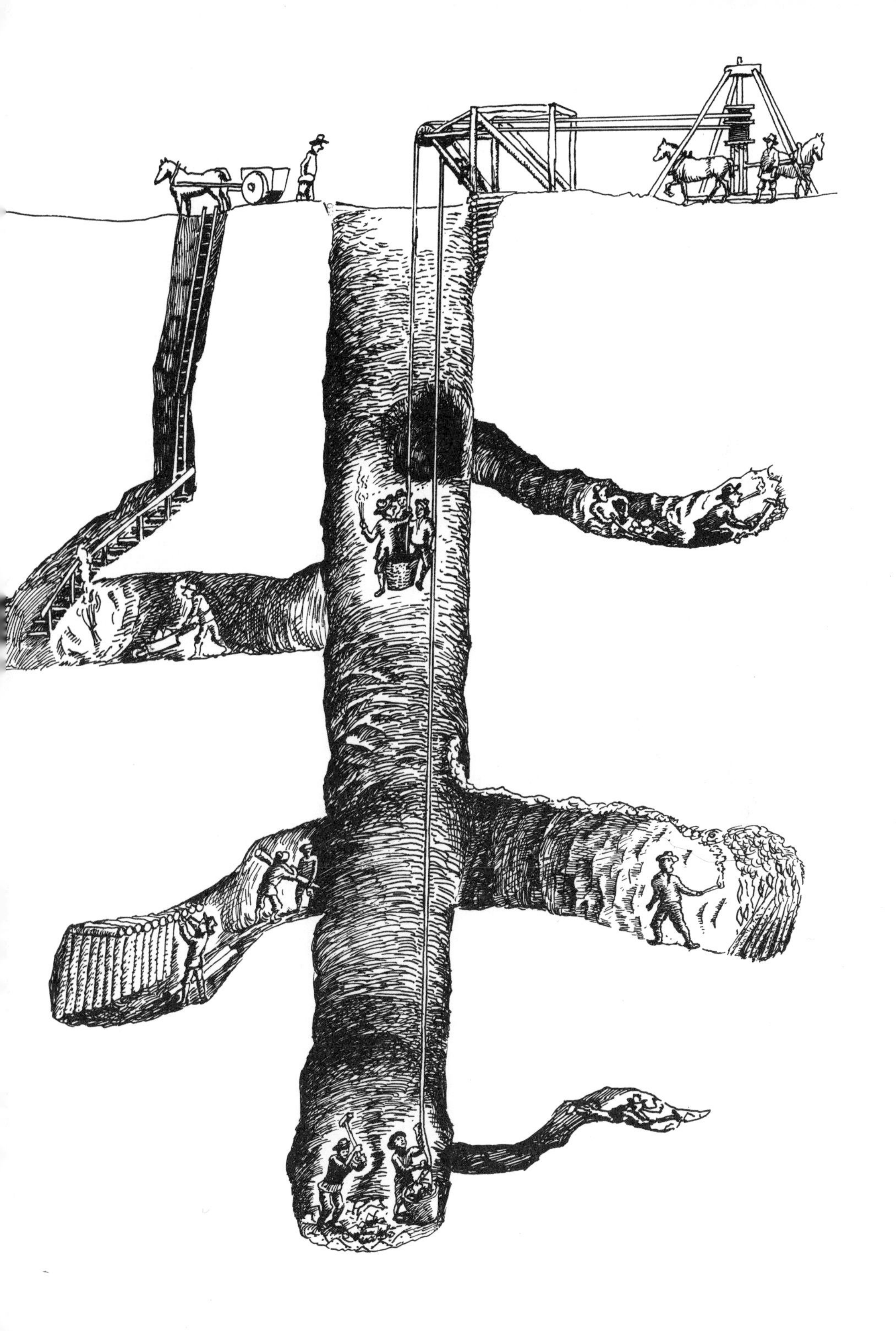

could result, should such a terrible and weighty construction collapse. For you who think you can see this in your mind's eye you will get, if not a complete picture, at least an impression of this highly strange and remarkable phenomenon". In the same context, Ogier also mentions that at the Copper Mountain, compared with other places, one could generally expect good, if uneven, veins of ore.

In contrast to earlier practice, efforts were made from the 1630s to develop the Mine in accordance with 'correct mining practice', following the Continental pattern.

An importent sign of this new orientation was the sinking of a number of large, broad shafts on the edge of the field. The first one was ready in 1648. By 1650, a further two shafts had already been started.

The Continental methods also included cutting galleries and tunnels, and building adits, which were now introduced at the Copper Mountain. In fact, not many galleries were constructed. It was easier to continue with the older chamber extraction method. On the other hand, more careful calculation and planning than before can be noted in the subsequent exploitation of the Mine.

The name 'adit' came to be used at the Copper Mountain for galleries which did not directly serve the mining process but were used for transport, ventilation and other similar puposes. However, these did not become particulary prevalent either.

The need for wood continued to increase and seemed enormous in the middle of the century. Mining now required up to 50,000 props, the equivalent of around 100,000 cubic meters, solid measure. In addition, a similar amount was also needed in the form of charcoal.

Furnaces

In the Mine, the Mine shareholders operated under the leadership of the Mine Master. The actual smelting of the copper, however, was the individual responsibility of each master miner.

Operations were carried on around the middle of the 1600s in over 130 furnaces and, as before, these were built at the different natural rapids found in the area, some with so little water that operations could only be carried out during a few months of the year. Several partners often shared one furnace, which was then exploited in proportion to the holding.

During the heyday of the 1600s, interest in improving the smelting process was almost non-existent. The furnaces, and with them the smelting method, were passed down from father to son and there was no reason to question tried and trusted technology.

However, one new item was introduced. From the 1620s, the expensive leather bellows could be replaced by bellows made of wood. These were originally invented around the middle of the previous century. At the Harz works it seems that wooden bellows had been introduced in the beginning of the 1620s and they were introduced into Sweden in the latter half of the same decade. This was thanks to Hans Steffens who was born close to the Harz Mountains and came to Sweden around 1625. He built the first wooden bellows for iron furnaces in southern Dalarna and introduced the same construction at the Copper Mountain.

The theory was simple. The bellows comprised two wooden boxes. The one underneath was fixed whilst the upper functioned as a moving lid. When the upper box was pressed down it forced the air out. In practice, a not insignificant problem was how to keep the construction airtight.

The first stage in ore processing, cold-roasting, was often carried out in pits situated near the Mine. Cold-roasting pits were greater in number than furnaces. When it came to turn-roasting, however, the master miners preferred to perform this activity close to their furnaces. Each miner had his own roasting house.

The smoke from these different operations, which of course consisted mostly of sulfur dioxide, was devastating to the vegetation. It was mostly the smoke from the cold roasters which gave the area around Falun its appearance, frequently mentioned, frightening and desolate.

The areas between the Mine and Falun river were the most exposed. This included the district of Elsborg to which the poor population, who had previously lived near the mines, was moved in the middle of the 1600s. The same applied to the meadowland down near the river, which ceased to be fit for cultivation at about the same time. The area was instead allocated for building purposes.

The smoke from roasting had become a feature of Falun. The Uppsala student, Andreas Julinus, explains on his travels: "The outside of the houses have become stained by the mass of smoke as if they had been covered with tar. And what is even stranger is that sometimes people on the street bump into each other since they cannot see each other in the middle of the day".

In a way it is striking in this context that perhaps the most knowledgeable description of the mining and smelting process, namely the one given in the Jars brothers' 'Voyages Metallurgiques' (from the 1760s) does not mention the smoke from roasting at all. For a professional it was obviously trivial.

Naturally, the smoke from roasting was an inconvenience but it was also considered a sign of human industry and inventiveness.

Not least, it was a confirmation of success and the prospect of earnings. It was not without reason that Queen Christina, who visited the Copper Mountain in 1649, expressed the hope that the smoke might never disappear.

Prosperity's men

In 1650, the Mine yielded the Crown almost 1,000 tons of raw copper, worth roughly 160,000 riksdaler. Though the sum was considerable, it nevertheless had only a limited effect on the affairs of the nation.

For some years around 1650 the country experienced a short period of peace. The government and the nobility were eager now to harvest the spoils of victory. Rewards were now to be reaped for all the trials and efforts of the war years. But the adjustment from provincial neck-of-the-woods to major European power could not take place without friction of some kind.

Economic problems piled up as a consequence of the soaring material and cultural ambitions of the monarchy and aristocracy. Newly-acquired resources seemed vast; demands and requirements were insatiable.

In the autumn of 1650 the coronation of Queen Christina was celebrated with a splendor previously unparalleled in the realm.

A procession marched through the capital preceded by heralds, kettle-drummers and trumpeters. Their instruments were of silver and they themselves were attired in velvet, bearing the Queen's coat of arms. They were followed by heads of the noble families of the realm, councillors and officials of state, walking or travelling in gilded coaches. The Queen herself travelled in a carriage drawn by six white, silver-shod horses with gold-embroidered velvet saddle-clothes. In Norrmalmstorg, the procession at one point passed under a triumphal Roman arch decorated with Latin inscriptions.

On the return trip from the Great Church to the castle, the treasurer walked in front of the Queen's carriage scattering gold and silver coins among the populace.

The days that followed were filled with a variety of public entertainments: ballets, allegorical tournaments, masquerade processions, tilting at the ring, pastoral plays. The text for one of the most splendid ballets was written by Sweden's foremost scholar of the time, Georg Stiernhielm. The visiting philosopher Descartes was also engaged to compose a similar work. The status of the royal Swedish 'salon' began to vie with that of its prototype in Paris. Uncouth Swedish soldiers were reeducated to suit the new standards of a chivalrous lifestyle.

Displays like these, however, did not satisfy Christina. She was anxious to demonstrate the resources and grandeur of a great power in the making by other means. Briefly, she wished her realm, and especially her court, to become a European center of culture and science.

One basic prerequisite for this lay in the rich treasures in the form of manuscripts, books, objets d'art and paintings that had been brought back from the theatres of war on the Continent. The Emperor Rudolph's collections had been transferred to Stockholm from Prague in 1649. The books alone filled 30 large boxes. The Queen also sent agents to buy manuscripts and books throughout the Continent. And foreign scholars flocked around Christina.

Uppsala University flourished. Ever since 1624, when it had been given a firm economic foundation by Gustavus Adolphus' magnificent donation of the gustavian hereditary estate, the university had an increasing number of professorships, by this time 20 in all. The library grew, thanks to the spoils of the Continental wars. Scholars of European standing could be found among the professors. The number of students burgeoned. There were probably 1,500 in 1650, and student clubs or 'nations' began to spring up. The Dalarna Nation seems to have been the first.

New universities were founded within the borders of the realm, in Dorpat in 1632, and in Åbo in 1640. In the diocesan towns, 'colleges' were established, dating from the 1620s, and these often maintained a high standard of teaching.

The predominant stream of thought in this period was neohumanism. In dissertations and orations, the students at the universities propounded the ideals of classical antiquity. Not even the orthodox priesthood, which was in principle very strict, remained unaffected.

At splendid festivities, pageants and ballets, participants appeared as Greek and Roman gods. The classical world view was not seen as the history of a distant past—its ideals had relevance to the current way of life, then and there.

Besides the classical world. Sweden had its own historical tradition, the gothic. The grandiose prehistory presented by Johannes Magnus in the middle of the 1500s, among others, had become something of a national ideology. On the basis of the gothic version of history, Sweden, considered an upstart among the European powers, could have pretensions to a past at least as glorious as that of the older civilizations. Scandinavia was the cradle of the races. The widely-read Georg Stiernhielm, after many years of philological studies, had come to the conclusion that Swedish was the oldest language.

Stiernhielm, who came from a mining family from the Falun area, was a multi-faceted author and official of state. The epic poem 'Hercules' not only establishes him as the portal figure of more recent Swedish literature—he was also a learned mathematician, a philosopher in the neoplatonic vein, and an archeologist.

The country also had prominent poets writing in Latin. Johannes Narssius, a Dutchman, was the country's state historiographer and described the feats of Gustavus Adolphus in heroic hexameters. Jonas Columbus, a cleric from Dalarna, also gained a certain reputation from producing similar work.

Runic learning was represented by Johannes Bureus. His cousin Andreas became the founder of Swedish cartography. His great six-page work on Scandinavia marks a Swedish breakthrough in the art. The map was drawn for military puposes and was not intended for general use. However, it began soon enough to be used as the basis for the work of cartographers on the Continent.

Olof Hansson Swart was one of Bureus' collaborators, and he played an active role when, at the end of the 1620s, Bureus was given the task of organising the Swedish land survey. Swart was the master craftsman behind the first map of the Mine in 1629.

Not only did royal favors shower over the representatives of science and the arts. Even greater gifts fell to the lot of the country's aristocracy. In the middle of the century, the Swedish high nobility enjoyed unparalleled political power, while fiefs and donations together with the imported booty of war gave them vast economic resources. The pomp which resulted was of a class to equal that known on the Continent. A retired

Swedish field-marshal could enjoy a life-style that did not significantly differ from a lesser German prince. Chanut, the French ambassador to Stockholm, thought that the Swedes were, of all peoples, the most enamored of pomp, considering their resources.

A man of the world was expected to have intellectual interests. Large libraries, widely varied in content, were often collected by the owners of newly-built castles. It is not entirely clear who actually read these books. Perhaps aristocrats who had fallen from favor had time for reading.

Christina did not merely wish to favor the leading noble families. She wished her court to be resplendent with highly titled men and, by an assiduous process of raising them to the nobility, she doubled the number of noble families during her reign, creating on average one new noble a week.

The 1600s have been called Sweden's international century. On a broader front than ever before, Swedes had come into contact during the wars with the milieux and people of the Continent.

The Swedes themselves, as opposed to mercenaries, also made a not insignificant contribution to the armies in the field. At the end of the war there were 60,000 men involved, and there was lively traffic between the homeland and the various theaters of war.

The number of Swedish students at foreign universities had grown considerably, and this also applied to the number of young nobles on their European 'grand tour'. As a country to visit, Germany, devastated by war, had been replaced by France, England and the Low Countries. A craftsmaster would never consider employing a journeyman before he had completed a long tour of work in foreign parts.

Coincidental with growth in Swedish travel abroad, foreign immigration to Sweden also reached a previously unheard-of level. There were soldiers from Scotland, smiths and craftsmen from the Low Countries, miners from the Harz. There were architects from France, and, not least, wealthy merchants of international standing from the Low Countries and England. These individuals—admittedly in return for substantial financial rewards—consistently made a great contribution to the country. At least four languages were spoken in Riddarhuset, the Swedish equivalent to Britain's House of Lords. The way in which the majority of these foreigners quickly acclimatised to their new country has often been emphasised. After only one generation, a De Besche could hesitate to undertake work in France because he was unacquainted with the language.

Colonial companies provide a characteristic insight into the Swedish spirit of the times. New Sweden on the Delaware River experienced its greatest prosperity during this period, under Lieutenant Colonel Johan Printz's governorship, between 1643 and 1653. The Swedes were more active in farming than in trade. But the company which led to the foundation of Cabo Corso (in present-day Ghana) was markedly commercial in nature , and here the plan was to buy slaves, gold and ivory. The African Company was founded in 1649, but its history stretched over fewer than 10 years.

The wish to promote development within the country led to the formation of towns, especially in peripheral areas. The objective was largely fiscal. The purpose was to establish routines designed to achieve better control of trade, the crafts and, of course, mining. Many newly-formed towns were not much more than villages. Most of them had fewer than 1,000 inhabitants. Following an initial period of establishment around 1600 and another around 1620, nine town charters were granted in the 1640s. As a result, Falun was joined by Kristinehamn, Vänersborg, Åmål, Askersund and Gränna.

Stockholm, the capital of this great power, developed considerably during the first half of the century. At the start of the century it was still a small town of fewer than 10,000 inhabitants, whose low houses had turf roofs and high whitewashed chimneys. In 1625 the town suffered a severe fire which gave the government an opportunity to regulate the old medieval street system, at least to some extent.

By the middle of the 1600s, the town had been transformed into a city fully comparable to the major urban centers on the Continent. Due to the national institution of fixed council administrations and central administrative boards, there was now a settled population of senior officials and state administrators. At the same time, the merchants had increased in number and become considerably wealthy. Stately new manors and palaces were the first thing to catch the eye of the visitor. Skeppsbron Avenue and Stora Nygatan were two of the capital's finest avenues where the buildings, which were almost entirely new, had characteristically high gables.

Dating from the first few decades of the seventeenth century, Sweden could with reason count itself among the most developed European countries with respect to public administration. One possible reason for this degree of sophistication may well have been freedom from any limiting traditions when the system was being established.

Central organs of government and new county councils were vital if the resources of the new, vast and sparsely-populated power were to be utilized efficiently. It was important to collect taxes and customs-and-excise duties, but it was also important to establish effective systems for monitoring activities within the armed forces and to maintain an accurate census of the population, especially of those able to bear arms. All resources had to be available for utilization in the state of crisis which was a permanent phenomenon.

Sweden had thus in many respects become a boom society, dynamic and open, in close contact with the major intellectual, economic and political centers of the world.

Losers

But parallel to this world of lavish luxury and cheerful industry was another, a world of distress and desperate poverty. Generally speaking, it was the Swedish peasant who bore the burden of maintaining the nation's greatness. There were many years of need. The year 1650 was one of them. This was the year when even people living in the most fertile regions were forced to eat bark bread. People starved to death, the roads crawled with beggars. It was a year in which the commoners of the Swedish parliament roundly attacked the nobility. Rebellion broke out in central Sweden. In connection with the festivities surrounding the coronation, the queen had ordered that all beggars who frequented the streets and alleys of the capital were to be 'removed', since they gave such a negative impression to visiting dignitaries.

For the majority of the population, Swedish living standards were actually lower during the nation's 'golden age' than in the sixteenth century. This is most clearly illustrated by studying the quality and quantity of the food eaten at the larger farms in various parts of the country. Such a study reveals that the standard of food at these farms was actually inferior to what was considered merely average a century earlier.

However, it appears that the Swedish Crown was the poorest of all.

Huge enfeoffments had drastically reduced state revenues. During the reign of Queen Christina, crown estates had been parceled off at a rate equivalent to 80,000 silver dalers a year. By the end of Christina's reign, more than 60 percent of agricultural land in Sweden-Finland had been transferred to the nobility.

It should also be noted that this comprehensive enfeoffment of land was not merely a symptom of Christina's extravagance, despite the fact that this was considerable. It was also the result of the state of the nation's treasury, which permitted no other form of recompense to those who returned victorious from the battlefield. The 'satisfaction', amounting to five million riksdalers, awarded to Sweden when peace was declared in 1648, was unimportant in this context and was in any case never paid in full. There were no new cities or nations to plunder.

Sweden had almost bankrupted itself recruiting soldiers at the beginning of the war. Due to other reasons, it proved equally costly to make peace.

The Crown was in a state of permanent crisis. There was no money to pay the salaries of government officials or other running expenses. The Crown was forced to borrow money to keep the system going and to draw advances on predicted cash revenues. Almost anything was used to secure these loans, including estimated sales of copper and incoming subsidies and customs dues, whether large or small, as well as any other means. To compound the problem, anticipated revenues were almost always overmortgaged. The result of this series of more or less improvised solutions was that a pile of outstanding debts were simply pushed forward, to be dealt with at some unspecified date in the future.

For those who were responsible for the fate of the nation, the international political situation must have constituted the greatest problem of all. The peace treaties signed during the 1640s were fragile constructions, the result of loosely formulated compromises and—as was the case with Denmark—only provisional. The empire was geographically far-flung and disunited. The old hopes of turning the Baltic into a Swedish inland sea remained unfulfilled. The new conquests, especially those in northern Germany, seemed more like unfinished initiatives. The neighbouring states ranged along the kingdom's long frontiers were mostly restless and hostile. Contemporary observers must have concluded that it was merely a matter of time before war broke out again.

Communications

After the various peace treaties concluded in the 1640s, reliable communications by land and sea with the kingdom's far-flung provinces acquired new and increased significance.

Based on the routes used for the royal progresses of medieval times, the national road network was extended during the 1600s, and the quality of the roads themselves was also improved.

The basic premises were far from promising. The roads which did exist were narrow, often had steep sections, were full of holes, and were broken up by roots and large stones which caused endless delays, hindering carts and wagons. Bridges were scarce, and travellers had to cross rivers by ferry or ford. In principle, the individual landowner was responsible for the upkeep of the sections of road which passed through his domain, although when a major enterprise was involved, such as the building of a bridge, one or more 'hundreds' (an early term for a rural district) were involved. Certain enterprises were large enough to be considered of national importance. One such was the widening and improvement of the trunk road which followed the coastline on both sides of the Gulf of Bothnia.

Out in the woods in the thinly populated northern parts of the country, the situation was different. There, the traveller crossed miles of marsh on a road which consisted of split or whole tree trunks laid end to end in the direction of travel, or over causeways where the logs were placed transversely, and packed tightly together. In the vast forests, the road was often marked simply by blazing the trees.

It was decided that a trunk road should be between five and seven meters wide. The actual road-construction process was of the simplest kind. On firm ground, it often involved no more than evening out the surface and, perhaps, laying gravel. Travellers normally walked or rode. Goods were transported by pack-horse. The wagons that did exist were simple and solid in design. The larger roads were constructed on the assumption that they would be used by various forms of horse-drawn conveyance. Carriages normally used suspension systems based on leather straps.

It was not only in the more centrally located areas that communications improved. There was intense activity in other areas as well. One example was Dalarna, which was important in terms of traffic. During this century, there was a gradual transition to wheeled transport. The many ferries came to be seen as an increasingly troublesome obstacle to smooth communications. As the century progressed, they were therefore

to a large extent replaced by pontoon bridges. Of these, the first was the 300-meter-long bridge over the Dalälven River at Brunnbäck, in the far south of the province, constructed in 1634, which was to be followed by a further 15 bridges. The bridge was so designed that it sank below the surface of the water when a heavy load was being transported across it.

A small number of foreign travellers noted their impressions of various trips made through the country. One of these was the English ambassador, Bulstrode Whitelocke, who made a trip at the beginning of the 1650s from Gothenburg to Uppsala with several carriages and a hundred smaller wagons. The trip was completed without too many loud complaints, which was probably due to the appalling condition of the roads in the ambassador's own country. Nevertheless, on his arrival at Uppsala, this elevated personage claimed that many of the major and dangerous campaigns in which he had been involved had proved less exhausting than this journey.

Twenty years later, the Italian diplomat Magalotti had nothing but praise for the Swedish roads: "There is no doubt that the traveller will find no better facilities for his comfort than in Sweden." Magalotti was particularly impressed by the number of carriages available and the "really quite comfortable" beds in inns along the way. According to a statute of 1649, guesthouses were to be located at twenty-kilometer intervals along the road. Since the price was determined by the number of kilometers travelled, the distance to the nearest surrounding villages was always displayed on a board at each staging post. This was also the reason for erecting milestones along the way.

The poor roads were one of the reasons travellers often preferred to go by sea, especially on long journeys. Yet this choice also had its inconveniences. Contrary winds could lead to weeks of delay. This applied especially to the Stockholm archipelago where, as one frustrated foreign traveller put it, you had to sail round almost all points of the compass before reaching port.

All the descriptions of travel in Sweden have so far referred to what was possible during the more clement seasons of the year. Conditions altered dramatically during hard winters, of which there were many during the 1600s, winters which brought snow and ice. Winter was actually preferred for the transport of heavy loads, such as ore, metals and timber. It was cited as something of a wonder that the transport of copper up and down the main road between Dalarna and Västerås continued on a year-round basis. The explanation was that it involved a raw material which was simply too valuable to store. The farmers bore the responsibility for ensuring the smooth flow of traffic during the winter months as well. This meant marking safe routes across the lakes to avoid the dangers of weak ice.

The water routes which criss-crossed the country were skillfully exploited. The idea of simplifying the transport of goods through the construction of canals and locks was obviously attractive. Some of these ideas were also actually implemented. One lock was completed on the Göta River in the early 1600s, upstream of present-day Gothenburg.

An enterprise which was to be particularly significant for the transport of iron was the construction of a canal between Lakes Mälaren and Hjälmaren. The first version of this canal was completed in 1639, although it was of little practical importance at this stage.

Post

In the 1600s, there was little or no integration between the various forms of transport available to the long-distance traveller. Things were even worse with respect to the transport of goods: the best thing to do was to accompany them yourself, or send a representative with them.

Traditionally, the common people had shouldered the burden of providing transport and lodging for travellers. However, as the volume of traffic increased, this became a very heavy burden, especially for those farmers with land which bordered the main highways. The serious difficulties with which the coaching services were struggling led to the establishment of an organized postal service. In addition, increasing military and economic contact with the Continent gave the Crown an urgent motive for establishing better communications, especially over long distances.

The first regular postal route was established between Stockholm and Hamburg in 1620, passing through Danish territory. Routes from Hamburg to virtually the whole of Europe were already in operation. The principle, which could not always be maintained, was that post should be despatched once a week. It was the responsibility of the county governor to ensure that fresh post horses were ready and waiting at thirty kilometer intervals.

Another initiative of quite different dimensions was launched on the instigation of the chancellor, Axel Oxenstierna, in 1624. This entailed the establishment of a "mounted postal service to ply all the main roads of the kingdom". Unlike the existing practice of using couriers, the idea was founded on a relay system. Postmasters were to be appointed to oversee operations. Individual letters were to be paid for according to weight and destination. These plans were finally realized in the Postal Act of 1636.

Initially, five postal routes were established. Apart from the original postal route via Denmark to the Continent, the routes were to Gothenburg, the Copper Mountain at Falun, to the north of Sweden and, finally, to Finland and the Baltic provinces. By the beginning of the 1640s, Sweden boasted a total of 60 post offices.

Given the geographical position of the country, the seaborne postal service was of considerable importance. As well as the regular seaborne postal service between the two large Baltic islands of Gotland and Öland, routes were also maintained across the sea to southwestern Finland and even further, to Estonia.

Shipbuilding

The Swedish merchant fleet developed rapidly during the 1600s. The largest shipyards were in Stockholm, Gothenburg and Västervik, on the east coast. However, skillful shipbuilders were also to be found in the eastern part of the kingdom (the western coast of present day Finland). The foremost military shipyard was in Stockholm.

During the early part of the seventeenth century, almost all vessels derived from Dutch designs, which meant a fairly rectangular midships section, the greatest breadth being at the waterline, and a slightly deep-drawing keel.

The largest warships, royal ships of the line could carry something in the region of 500 'lasts', a 'last' being a unit of measurement of weight for a ship's cargo, a measurement which varied with time, but which was normally slightly in excess of a ton. The shipyards also designed and built men-of-war to carry as little as 200 'lasts'.

The armament was steadily increased. The royal flagship Wasa—famous because it capsized on its maiden voyage, got salvaged and is now on display—carried 64 guns.

Iron canon—originally cast, later forged—gradually replaced the earlier bronze canon. But Wasa was still equipped with bronze canon, despite their higher cost. The reason for this was their light weight compared to iron canon. The sad fate of the Wasa in 1628 did not halt the construction of large ships. Over the next eight years no less than five new ships were built, four of them larger than Wasa.

During the first part of the seventeenth century, the master shipbuilders worked according to the old traditions. The basis for the ship's design was an 'evaluation', which normally involved some fairly approximate data: the required length, breadth, height and other key dimensions. Sometimes a model was built, which could be divided into separate sections.

Fredrik Henrik af Chapman, the famous ship designer of the eighteenth century, has described the situation in uncompromising terms. According to him, foreign shipbuilders, in the true tradition of craftsmen, designed and built their ships on highly individual concepts—which is why vessels designed to carry the same cargo varied so greatly in size. When the canon were ordered, it was the captain of the ship who determined their size and weight. The rigging master decided what masts and rigging were to be used. Each individual developed his own ideas. In Chapman's view, ships built in the early part of the 1600s were often far too small and fragile for the size and weight of the armament they bore.

The forest industries

As late as the mid-1600s, Sweden was still, in industrial terms, an undeveloped country. With respect to international trade, the nation's natural role remained that of the raw material supplier. The nation's iron and other metal resources were the determining factor.

The basis for all metal processing were the nation's forests. Only about one percent of the country was being cultivated. Sweden was a land of forests. Apart from private woodlands, the forest remained a free resource for the entire community for many years to come.

The forest provided raw materials for much that was essential to the daily life of a seventeenth century person. The forest provided timber for houses, furniture, domestic implements and fences.

Felling was always carried out with an axe. Using this technique could result in stumps which were as much as a meter high. The forest also yielded birch-bark, which provided an efficient intermediate layer of insulation when laying roofs. Pine bark was sometimes used as feed for cattle in the early spring, when winter stocks of hay were running low. It was 'skinned', cut into strips and dried. The bark was rich in vitamin C and helped to prevent illness related to vitamin deficiency. The poor often mixed bark into their bread, even during normal years.

Tar was a vital commodity. It was used as a lubricant and as a protective covering on roofs and on all types of vessel. It was also used in medicine. It was exported in considerable quantities. During the 1600s, the Swedish-Finnish Empire had a virtual European monopoly on tar production.

Pitch, the glutinous substance which remained after the more volatile substances had been boiled off, was also exported. Pitch was in great demand as a sealant, one of its applications being to caulk wooden hulls below the waterline.

The annual volume of tar and pitch exported from Sweden suggests a consumption of almost 2 million trees.

Potash (potassium carbonate) was another product that was marketed, used in the glass industry, but also in the production of lye and soap. The raw material on which potash was based was hardwood. The dried wood was piled onto a great fire. The ash was then collected and soaked in water. After this, the water was evaporated by boiling the mixture in pots. What was left was then annealed. It took 1.5 cubic meters to make one kilogram of potash.

The three products described here are perhaps the most extreme examples of how forest resources were commonly wasted at this time, but it also applied to other products. More than a meter-long section of tree trunk was used to make a bowl: the rest of the trunk was left to rot.

When planks were needed to construct a house, logs were trimmed from both sides until a plank of the right thickness was achieved. This method yielded a single trimmed plank per log. Logs were also split using wedges, which at least yielded two 'planks' per log. These types of split log could be used as flooring, with the flat side uppermost. In northern Sweden, the practice of producing 'trimmed' planks continued right into the nineteenth century.

Sawing logs consumed almost as much timber as the above methods. The older saws had cutting blades which could be up to a centimeter thick. Not only this: they were also difficult to use.

While the Crown generally restricted the activities of the sawmills, it encouraged another activity, unconnected with the common people's direct means of earning a living, yet which also consumed large volumes of timber: charcoal burning in the forests. For many of the peasants and small farmers in the mining districts, charcoaling for the open-hearth furnaces and hammer forges was a primary occupation during late fall.

And then came timber-guzzling mining operations. In the mid-1600s, mining and metal processing were leaving their mark on large tracts of forest in central Sweden.

In the context of these many and, each in their own way urgent needs, it was natural that a debate should arise about the threat of a possible scarcity of timber. This debate really gained momentum during the 1630s, resulting in the drafting of a Forest Statute in 1647, the first of its kind in the world. It contained a nationwide ban on the burning of woodland, unless its purpose was to clear land for permanent use as arable or pasture. The timber forests and trees such as oak, hazel and beech were awarded special protection. Further legislation was introduced, which sometimes involved extremely detailed instructions. The right of the individual to decide what could be done with his own forest was drastically reduced.

However, there was never any scarcity of timber in the true meaning of the phrase. On the other hand, it is clear that conveniently located areas of forest were ruthlessly exploited. The real problem was transport. The great tracts of forest remained untouched. The fact that contemporary commentators had a vague and unreliable grasp both of the true extent of the nation's forest resources and of the forest's ability to grow back led to an unjustifiably pessimistic evaluation of the situation.

Iron

During the latter half of the sixteenth century, Sweden doubled its iron exports and, by the beginning of the seventeenth century, they exceeded 8,000 tons a year. The subsequent pace of development accelerated, and towards the end of the century, exports were close to 30,000 tons a year. At the same time, the quality and appearance of the product also changed.

Around the turn of the seventeenth century, exports still consisted largely of osmund iron. By 1640, the situation had changed. Osmund iron had vanished from the statistics, and had been replaced by bar iron. Technically, the difference was that iron processed in a forge was stretched under a water-powered hammer, while osmund iron was worked by hand.

The raw material thus processed came from the widespread iron ore deposits in central Sweden. As late as the mid-seventeenth century, mining was still a seasonal activity, although at the sites of the largest deposits, a new social class was developing: that of the professional miner.

At this date, mines were still not worked at great depths. At the beginning of the century, the Norborg deposit had no mines deeper than eight meters (approx. 20 feet). In other mines, depths of between 30 and 40 meters were common. Due to the fact that these mines penetrated well beneath groundwater level, some form of pump had to be used, operated either by hand and a treadwheel, a waterwheel, or by horse-powered winches.

Operations based on the rich Dannemora deposit in Uppland, north of Stockholm, developed into a sizeable enterprise due to the rapid growth of the ironworks in the province, some owned by immigrants from the Low Countries (Walloons), others by Swedes. As early as 1610, Dannemora was supplying seven different ironworks. By mid century, production was close to 10,000 tons of ore-bearing rock a year. At this time, ore was being worked at four different mines located on the same deposit. One of these mines is thought to have reached a depth of 60 meters. At this point in time, 12 mining companies had been granted mining rights. The work was led by privately appointed mine bailiffs, who concentrated primarily on furthering the interests of their employers. This therefore made it difficult to achieve any really efficient planning of mining operations.

As the century progressed, the traditional Swedish blast-furnace faced competition from two new types of furnace. One of these, the German furnace, was in principle constructed in the same way as the older Swedish version, but was larger and more solidly built. The upper section of the German design comprised rock fill, held together by timber walls. The dam was constructed in stone or iron, and was sealed with cinders. The slag was tapped in batches through a special hole. It was these earth-and-timber furnaces, which were fairly easy to build, that were normally used by the peasant miners of the period.

The other new type of blast-furnace was used by the Walloons, and was known as the 'French' blast-furnace. It had solid stone foundations. The outer sheath was also constructed in stone. This type of furnace was also taller than the other two older designs, standing more than eight meters above the ground. In the 'French' blast-furnace, the slag was allowed to pour over the dam continuously. Relatively expensive to erect, these 'French' blast-furnaces were mainly found at ironworks.

By the middle of the century, a peasant miner's furnace produced in the region of 30 tons a year, while the 'French' ones produced perhaps four times as much.

Forging

Compared with a blast-furnace, a hammer forge was a large and expensive installation which required a considerable volume of water for power. From the beginning of the seventeenth century, therefore, almost the only operators with sufficient capital resources to erect such installations were the works owners, while peasant miners had to content themselves largely with the production of pig iron.

The two main methods were German forging and Walloon forging, the German method being the most common. This method permitted the use of the same hearth, both for decarburization of pig iron (a process in which elements which make iron unforgeable—in particular, carbon and silicon—are removed) and for the next stage of the process, forge welding, in which the pieces of iron were reheated and subsequently compressed under the hammer. In general, hearth decarburization produced a slag-rich iron characterized by uneven composition and hardness. In a Walloon forge, however, decarburization of pig iron was carried out on a special hearth, and forge welding was carried out on another. The product generated on this second hearth was a doughy mass of low-carbon iron, which was compressed to form 'blooms'. Once forge welding was completed, these were then forged into finished bars.

The Walloon forge was introduced during the 1640s at large ironworks in northern Uppland. The method continued to be used right into the twentieth century, with only marginal changes in technology. The iron produced by the Walloon method was hard, due to its high manganese content, and was surprisingly free from impurities, especially phosphorous. However, it was also expensive, due to a high degree of burnoff and high charcoal consumption.

In Sweden, a fairly unique quality assurance system was established. In the 1637 statute relating to hammer forges, the owner of a hammer forge was obliged to stamp his iron so that suppliers of substandard iron could be traced more easily. This stamp became a symbol of quality.

Weapons

Although the international role which Sweden played during the 1600s was that of an exporter of raw materials and semi-finished goods, there was one significant exception: weapons.

The most important manufacturing center in olden times was Arboga, which started to develop around 1550. By the 1620s, fifty craftsmen were permanently engaged in plying almost 20 different specialist trades. This involved more than the manufacture of firearms. The production of weapons such as swords, halberds and pikes accounted for a significant part of this industry. Harness was also an important product.

Nevertheless, weapons production was by no means restricted to Arboga. Rural smithies also accounted for a significant proportion of the country's total weapons production. Arms workshops were established in several towns. Even the manufacture of pistols was a decentralized activity, at least in the early stages. Subsequently, this form of manufacture was increasingly taken over by major entrepreneurs such as Louis de Geer. During the Thirty-Years' War, Sweden became self-sufficient in all forms of weaponry.

The musket, the most important handheld firearm, was developed in increasingly lighter versions. In 1646, the average weight of a 'slow-match' musket was slightly less than 5 kilos. Towards the end of the 1640s, production exceeded 10,000 muskets a year. Swedish weaponry compared favorably with Continental craftsmanship in this field.

As for canon, Sweden developed into one of the largest European manufacturers within a matter of a mere two decades, dating from 1618, exporting large numbers, which could compete with the best on the market, such as those made in England. The honor for such an achievement must to a great extent belong to Louis de Geer, who played a central role as a manufacturer and supplier of weapons during this period.

A key reason for this success was that Swedish foundry charcoal pig was considered excellent, and possessed unusually high durability and toughness. Nine standard calibers were available.

Iron canon competed with copper canon from Falun. Falun 'gunmetal' contained 90 percent copper and 10 percent tin. The tin made the metal harder but also made it more brittle.

When Gustav Adolf established the guidelines for effective Swedish field artillery, a primary objective was to reduce the weight of canon.

With this in mind, he introduced what were to be known as 'leather canon', constructed of gunmetal strengthened by iron collars and several layers of cordage and leather. They were extremely light, totaling 38 kilos, including the gun-carriage. In practice, they soon proved themselves ineffective, as they became rapidly overheated. When the Polish campaign ended in 1629, this type of canon was removed from service.

Gustav Adolf achieved much greater success with the small bronze or cast-iron canon designed for him by the eminent Colonel Hans von Siegrooth. These canon weighed 150 kilos, excluding gun-carriage, and could be drawn by one or two horses, and operated by two gunners. These regimental pieces of artillery became particularly famous, and were in great demand on the international market.

The Falun gun-foundry, which supplied a notable number of such artillery pieces, was granted special privileges in 1630, and remained active until the end of the war in 1648. Right from the very beginning, the foundry cooperated closely with Siegrooth. Between 1634 and 1642, the Falun foundry was probably the largest of its kind in the country, supplying a total of more than 1,000 pieces of artillery of varying calibers.

Silver

The exaggerated belief in the importance of mining precious metals (especially gold and silver)—at almost any price—continued both on the Continent and in Sweden.

Those in power thus made great efforts to develop operations at Sala.

These efforts were not without success. An increase in production *was* achieved in the 1640s, although it was still less than half the volume obtained a hundred years earlier. At the same time, the international price of silver was in decline. The annual yields reported during the 1600s never even reached a value of 100,000 silver dalers.

Much of the credit for this relative comeback must be given to George

Grissbach, a German immigrant who became one of the country's leading master miners. As well as being head of mining operations at Sala, a post which he held until his death in 1651, he was in 1630 also appointed Senior Master Miner, responsible for all mining of precious metals throughout the kingdom.

Operations at Sala were improved by sinking shafts and introducing water-powered pumps.

Before the new works could be built, something had to be done to improve the supply of water required for power at the mine. The water-control measures introduced by Grissbach were complex, necessitated by the flatness of the terrain, and involved a widespread system of dams and canals. More than 10,000 acres had to be covered by water.

Actual mining operations revealed no especially original thinking. Miners of the time were aware that working silver ore involved special dangers. The job of quenching the ore-heating fires was, with good reason, considered dangerous to health. The smoke from these fires contained elements such as lead, sulfur and arsenic. The mineworkers who carried out this work were paid more than the others, since "they were unlikely to enjoy such a long life". Conditions close to the silver smelters were just as bad. The mineworkers employed here rapidly became "shaky, red-eyed and sickly".

Ogier provides a highly concrete description of activities at the Sala mine in the two reports mentioned earlier. His comparison between the Sala deposit and the Copper Mountain is not to the former's advantage. "The difference between the silver mine and the copper mines is that the latter are much more beautiful, much richer, not so deep and have more workers than the former, where all one sees is horrifying depths, terrible descents and ascents, dangerous tunnels, few workers, few rich veins, and few fires."

According to Ogier, the silver mine reached a depth of 100 meters. The open-cast pit is thought to have been almost 90 meters across.

Although Sala was always a Crown favorite, this did not diminish enthusiasm for optimizing other silver deposits. This applied to Öster Silvberg, which hardly lived up to expectations, and Hällefors, which achieved real significance at the end of the 1630s.

The famous Nasafjäll mine in arctic Lappland was a doomed (although in its own way heroic) attempt to extract the riches believed to be hidden within the mountain, despite frightful conditions. Over the 23 years up to 1660 during which the mine was worked, no more than 36 kilos of metal was extracted on average each year. An equally unsuccessful silverworks was to be established in the same area at a later date. It was in these thinly wooded regions that one very special technical innovation—blasting rock with gunpowder—was to be tested for the very first time in Sweden. The method was first used in Germany in 1613, and at Nasafjäll a mere 20 years later. A single barrel, containing 0.165 cubic meters of gunpowder, was said to produce the same effect as gained from burning 1,500 cubic meters of firewood.

Once the practice of blasting with gunpowder had started, knowledge of the technique spread rapidly throughout the country, although it was applied more gradually. The traditional technique of setting ore-heating fires still offered certain advantages. The effects were easier to control, and produced thinner, conveniently sized slabs which were ideal for handling. Although black gunpowder was not a particularly powerful explosive, it was still difficult to anticipate the effect of blasting. It was also a new technique, which required new operating methods, such as drilling.

The men of the Copper Mountain

The Copper Mountain was also considered a precious-metal mine, although less elevated than the silver mines. In purely economic terms, any comparison was heavily weighted in favor of the Copper Mountain. Calculated over the century as a whole, national copper production, in terms of revenues, exceeded silver production by a factor of 13 or 14.

It has been stated that the workforce at the Mine up to 1650 totaled 400 miners or more. This figure is uncertain, since it refers to two different categories of miner. It comprises both workers employed by the central administration and those hired by individual master miners. It is difficult to assess the exact number of these latter workers.

In addition to directly providing work and sustenance for perhaps as many as 1,000 persons (master miners, their workers and families), the Mine also provided large numbers in the town of Falun with a living.

Even at the time the town of Falun was granted its letter of privileges, it was the second most important community in the kingdom. It was therefore entirely natural that one of the country's first postal routes should end there.

What the authorities in Stockholm had in mind was to turn Falun into a real mining town in the true Continental tradition. In Axel Oxenstierna's personal report concerning a town charter, which he wrote as early as 1618, he describes the town of the Copper Mountain as a community in which the master miners should play the decisive role. At this time, however, the miners, who already enjoyed wide-ranging privileges, showed only limited interest in the idea, and plans were dropped for the time being. However, after the dissolution of what had in any case proved to be a fairly unsuccessful copper monopoly in 1639, the original plans became of interest again, especially to the men of the Royal College of Mines. What the Mountain needed was an active class of burghers, and a town constitution. The town charter granted in 1641 was actually only intended as an interim measure, and Falun never really achieved the status of a formally constituted mining town during the seventeenth century. The cooperation which the administrators in Stockholm had hoped would develop between the active master miners and trading burghers remained, at least for the time being, a dream. The only thing that remained to reflect this dream was the town seal, which shows a stylized picture of the town with its rows of houses and, above them, separated by a flowing river, the flaming mountain (symbolizing its ore-bearing capacity), decorated with the copper symbol.

One of those who was certainly committed to thinking big about the future of Falun was Johan Berndes, the county governor, who had long held a commanding position in the Royal College of Mines and was well placed when the question of granting Falun a town charter became topical. When he was appointed governor in 1641, he found himself responsible for a large, newly formed county which comprised three old counties: Kopparberg, Salberget and Näsgård. Berndes was thus a government appointee, there to serve the government's interests at the Copper Mountain, at Sala (with its silver) and at Norborg (with its large iron mine). During the first few years of his appointment, he spent most of his time in Sala.

Berndes has been described as typical of the inventive, capable and multi-skilled officials who characterized Sweden during its period of greatness. When mining processes were being discussed, his opinion weighed as much as that of Karl Bonde, president of the Royal College of Mines, or Louis de Geer. It was he who initiated work on the first great shaft at the Falun mine, as well as the establishment of the Wood Company, which was to supply the Mountain, and a grain warehouse. A similar warehouse was erected in Sala.

Berndes was also involved in the establishment of a hospital and a school, not to mention the large church in Falun, which was named after the Queen. This was not all, however. He also had duties at a national level. Following the end of the war with Denmark in 1645, it was Berndes who received the province of Jämtland on behalf of the Swedish Crown from the Norwegian authorities, and who subsequently received the oath of allegiance from the people. After his term in Dalarna, he was appointed president of the National Chamber of Commerce, a member of the Privy Council and Governor of Stockholm. He is described as a charming personality who "seemed easy to approach".

By the middle of the seventeenth century, mining operations at the

Mountain were under the direct control of perhaps the most notable Mine Master of them all.

Hans Filip Lybecker was born in Germany, where he is said to have studied mine engineering under George Grissbach before, at the age of 15, in 1623, he travelled to Sweden. Once in Sweden, his career developed rapidly, and by 1644 he became responsible for mining operations at the Copper Mountain while retaining several other duties, which included responsibility for operations at Nasafjäll.

As a mining expert, he had no equal in Sweden, and he controlled the large Mine with a firm hand.

In addition to his everyday duties, he also found time to introduce a whole series of innovations. Several of these were related to the supply of raw materials for the Mine and the town. He was deeply involved in the above mentioned Wood Company, which was formed at the end of the 1640s, and which was responsible for organizing the substantial supplies of wood which came from the forests of western Dalarna. The Wood Company acquired an organizational structure which was to function efficiently for many years to come. Lybecker also took an active part in the purchasing of grain, a vital resource in a region of such scarcity as that surrounding the Copper Mountain. In the fall of 1650, he was authorized to purchase several thousand barrels of grain in Stockholm. The background to this was an unusually bad harvest. But he also took the opportunity to purchase large quantities of other commodities on behalf of the master miners.

As Mine Master, Lybecker was not allowed to own any shares in the Mine himself. However, he *was* allowed to participate in actual operations where new pits were being opened up beyond the perimeter of the central deposit, an opportunity of which he took full advantage.

The many interests the Mine Master cultivated that were incidental to his main duties, and the fat profits he was therefore able to make for himself inevitably led to criticism. He was clearly both versatile and extremely wilful, and often had difficulty in providing a satisfactory account of all the transactions in which he was involved.

However, it is clear that his sharp mind and constant inventiveness actively contributed to the tremendous development achieved by the Copper Mountain in the mid 1600s.

Refining

Crude copper had traditionally been 'refined' in the Falun region, to a limited extent, for use by the village smith in the production of various items. In principle, however, the Copper Mountain's end product was black crude copper. By 1650, the process, known as 'refining', had already been in operation at Avesta for ten years. Operations at Avesta were led by Marcus Kock, a strong-willed and knowledgeable engineer. Born in Liége, he had since 1613 been employed as Mint Master at two Continental mints before travelling to Sweden, where he became director of the Mint in Nyköping in 1627. When he moved to take over at Avesta, he was not only responsible for minting the realm's coinage: he also became responsible for the 'refining' process.

Transport of the copper from the Copper Scale in Falun to Avesta was handled by foremen who were responsible for ensuring that their specific load of copper arrived at the refining works within ten days of being weighed on the Copper Scale. This job of transportation was carried out at the foreman's own risk.

During the summer, the route led over Lake Runn to its southern shores, where the copper was reloaded for transport by road. During winter, the entire journey was naturally made by sledge.

The refining works could make a profit due to the fact that incoming crude copper was calculated in ship pounds (Mountain measure)—about 150 kilos—while refined copper was calculated in ship pounds (Stockholm measure)—about 136 kilos. The difference was not all profit: the refining process involved considerable burnoff.

More than any other installation in the Sweden of this period, it was the Avesta copper refinery which came closest to being a true large-scale industrial enterprise. The building in which the refinery was housed was about 75 meters long, 40 meters wide and contained six twin furnaces. In addition to these there was a twin 'rake' furnace for re-smelting operations, two hammer forges and storage rooms. The overshot waterwheels which powered the bellows for the furnaces and the hammer forges were fed by a broad conduit which passed right through the building and which divided it into two from one end to the other.

The Avesta furnaces were fairly large, with domed roofs and high chimneys. Each furnace with its two hearths formed a separate workshop, superintended by a special 'Refining Master'. Normally, not more than three plates of crude copper were smelted at one time, comprising a total weight of two tons at the most.

When smelting was complete (a process which took about 12 hours), it was time for the 'abrading' process. A scoopful of cold water was poured over the molten copper, causing the uppermost film of copper to harden into a sheet, which could then be drawn out of the furnace and submerged in a vat of water where it was quenched. This method was used to draw sheet after sheet out of the furnace: they could then be stacked in a pile. It was possible to get between 80 and 90 sheets of refined copper from a fully-charged refining hearth.

The Avesta end product was very pure copper (97 to 99 percent pure), and was well suited to the production of brass. It was, however, too brittle for forging, having a low cuprous oxide (Cu_2O) content. It could only be used for forging if submitted to a further smelting process, using charcoal.

In connection with the fact that the production of refined copper had reached adequate volumes in the country, a total ban on the export of crude copper was imposed from the beginning of 1625.

Refined copper was basically a finished product. Nevertheless, a fairly significant amount was further processed at the same location, in the Mint, which had been established simultaneously with the startup of the refinery. From 1644 onwards, this primarily involved the minting of copperplate currency.

The copperplate coins were forged in large rectangular sections. The most difficult aspect of this manufacturing process was to ensure that these coins contained the prescribed amount of copper. The round coins of lower denominations, however, were stamped out of narrow smooth-rolled copper bar. Smooth rolling had been used in Sweden since the end of the sixteenth century. However, it is remarkable that the sort of precision work involved in minting new coins could be carried out between rollers. The method was introduced by Marcus Kock. Nevertheless, this did not prevent the more primitive technique, the use of a sledge hammer, from being reintroduced at the beginning of the eighteenth century.

Copper on the market

The boom in the Swedish copper-mining industry occurred at a particularly fortunate time with respect to the country's trading position.

The quantities of copper produced around 1650, although they may appear modest compared to present-day volumes, probably corresponded to about two-thirds of total European copper production at that time, giving the country a near monopoly. The yield from traditional and older mining regions in Hungary, Bohemia and Thuringia had declined towards the end of the sixteenth century.

Copper prices had risen sharply since the beginning of the century. One key factor was Spain's surprizing decision to convert to a currency standard based on 'vellon', which comprised almost pure copper. This highly satisfactory situation continued for almost a quarter of a century, until 1626.

The slump which the copper market then experienced created problems for the Swedish state, especially in view of the fact that Sweden's participation in the Thirty Years' War was to a large extent funded by revenues generated from the export of Falun copper.

The question *now* was—how could the nation's commanding position on the copper market be exploited to the greatest advantage? Supplies of Japanese copper were not to become a threat until later in the century. The most immediate solution would be to cut back production, reduce the amount of copper on the market, and thus halt the decline in prices. However, this prescription was far from easy for a nation which found itself in permanent need of ready cash. The ideal was to sell as large volumes as possible—and *still* maintain high prices. Louis de Geer, who was a trusted intermediary, was able to achieve this economic rope-trick several times, by personally purchasing small quantities of copper at high prices, naturally making a loss, yet at the same time managing to spread the impression that there was a scarcity of copper. This type of strategy could only function for a limited time.

It was against this background that the idea of using copper for minting coinage was first raised. The original intention was not to manipulate the price of copper. Quite simply, the realm was suffering from a lack of silver. Copper was used to balance this scarcity. The minting of low-denomination copper coins first started in 1624.

However, the idea gradually crystallized that the minting of copper coins, on a sufficiently large scale, could also prove an effective instrument in steering the price of copper. When the international price was low (i.e. that quoted in Amsterdam), copper production could be absorbed within the nation's borders. When the price of copper rose again, minting of copper coins could be stopped or restricted.

One factor which contributed to this interest in minting large copper coins was that they were cheaper to produce than smaller denominations, where approximately 20 percent of the nominal value comprised stamping costs.

The first copperplate coins came into circulation in 1644. They were also the largest, weighing almost 20 kilos, and with a nominal value equivalent to ten silver dalers. These copperplate coins were intended to reflect actual value, and were therefore almost a hundred times heavier than their equivalent in silver.

Axel Oxenstierna's statement that these coins were 'portable', often described as a joke, actually contained a serious insight. Although large, these copperplate coins were still small enough to be transported by road to Gothenburg for subsequent export, and therefore did not have to be transported by sea through the Sound between Denmark and Sweden, which could be dangerous during times of unrest.

These copperplate coins soon vanished from the market. Of the 26,774 minted in 1644, and to some extent in 1645, no more than eight are still extant. The reason for this is that virtually all of them were soon sold as copper plate.

Actually, this development was not unexpected. From the very start, when the minting of copperplate was introduced, the Privy Council was well aware of the fact that these coins could be used both as currency and as traded goods. What apparently had *not* been foreseen was the subsequent degree of speculation in copper coin. Private buyers were able to change their funds into copper coinage at below-market prices, enabling them to compete successfully with the Swedish Crown on the Amsterdam stock exchange.

These copperplate coins were quoted as being of a special grade. They were fairly impractical in some respects, but they were ideal in one, albeit a rather specialized one: they were perfect as copper matrices for casting type. At least, this was the opinion of the great typographer Pierre Simon Fournier. The coin was of precisely the right thickness—all you had to do was cut it.

The basic objective—to control copper prices by varying the amount of coins minted—remained unfulfilled. Sweden's national economy was too weak to carry it through.

The minting of copperplate coins developed on a grand scale. On average, a quarter of all copper produced went to the Avesta Mint between 1624 and 1691. However, the number of coins minted varied greatly from year to year: in 1644 and 1645, almost half the total copper production was minted, in 1649 a mere quarter, and in 1650 as little as five percent.

The copper could also be used in a number of other ways. The canon foundry in Falun has already been mentioned. Copper was also promoted as a roofing material with some success. Many foreign visitors considered especially noteworthy the fact that many churches and large buildings had copper roofs.

From the start of the seventeenth century, increasing quantities of copper were being used by brassworks. Brass had been manufactured in Sweden for domestic consumption since the early years of Gustav Vasa's reign, especially in connection with the manufacture of weapons. True brassworks were established during the early years of the seventeenth century in Nyköping and Skultuna. Dating from the 1620s, the state authorities encouraged brass manufacture as a way of increasing the demand for copper.

It was therefore no coincidence that a brassworks, which was to become the largest in Sweden, was started in Norrköping at this time as the result of one of Louis de Geer's many initiatives.

Large-scale manufacture of brass, like the production of refined copper, was to a great extent a technical novelty in Sweden. As a result, Swedish brassworks were, in their infancy, led exclusively by foreigners. Many of these came from the famous home of the brass-manufacturing industry in the neighborhood of Aachen.

Brass was almost exclusively an export commodity. Almost 90 percent of production regularly went to export. Following a period characterized by falling prices in the 1640s, the economy improved and the Swedish brass-manufacturing industry really took off. The most important product at this time was wire.

Ogier visited the Norrköping brassworks in 1634: "It is amazing," he writes, "to see the huge water-powered forge hammers, as they beat and stretch the copper sheets with such a tremendous clamor that even Jupiter himself, God of Thunder, could not bear to listen to it on his heavenly throne, and to see the workers as they lay these sheets on the anvil to beat them with hammers or cut them with steel-shears into long strips, which they bear to another workshop where they are transformed into the wire we know as 'fil d'archal' or 'fil de richal'(brass wire), by being wound rapidly round rotating wheels by other fellow workers. And all this is carried out in the midst of such an intense, ear-battering squealing and screeching that one is convinced that even if the works were to be visited by the deaf, they would surely hear it."

Other products were thin sheet, supplied in rolls, and bases for pots.

A considerable amount of this production was supplied to Swedish craftsmen. The girdlemakers, who among other things supplied the army with buttons made from these sheets, were typical. Part of the semi-finished export product returned to Sweden in the form of finished goods. It seemed a good idea to try and carry out this final stage in the country as well. One of the results of this realization was the special-manufacturing operation at Skultuna, which had grown to quite a size by 1640, at which time the Privy Council was shown a "heap of different types of pins". This manufacturing activity expanded. Pin makers remained active into the nineteenth century, when their product could no longer compete with the factory-produced pins being made in England. Other finished products manufactured in Sweden included pots and chandeliers (a Skultuna specialty) and similar items. To these were added "many ornaments used in the home, for carriages and on harness".

Around the middle of the seventeenth century, exports of refined copper were still seven times greater than exports of brass. Three decades later, they were equal.

The manufacture of brass involved the use of galmey, or zinc carbonate ($ZnCO_3$), due to the fact that zinc could still not be produced in a metallic form. Galmey and copper were melted together in a crucible, using charcoal. The alloy normally contained approximately two parts copper to one part galmey. The galmey had to be imported.

It has often been stated that the Copper Mountain's period of greatness, like that of the realm, was short-lived, and that the highpoints of both more or less coincided around the middle of the seventeenth century.

The copper-extraction industry declined rapidly, dating from the beginning of the 1680s. Swedish trade statistics indicate that copper and iron jointly accounted for approximately 80 percent of total exports for the century as a whole. The domestic situation changed dramatically, however, and the relative importance of copper declined as that of iron rose.

If one postulates the thesis that Sweden's position as a major power could only be maintained through permanent overstretching of the nation's resources, the story of the Copper Mountain provides excellent support. The major and catastrophic cave-ins at the Mine in 1655 and 1687 were clear omens.

At the same time, it may be said that the maintenance of the nation's status as a major power, despite the crippling failures which led to the loss of this status, nevertheless created a pool of know-how, administrative expertise and scientific learning which was to prove invaluable during the next and less brilliant stages of its development.

The practical theoreticians

IN THE 1770S, IRON accounted for 85 percent of the Swedish mining industry's total production, while copper accounted for only 12 percent. Although this figure was high enough to make copper the country's second most important export commodity, at the same time, it was an indication that the significance of copper for the national economy had drastically diminished.

The situation at the Mountain was difficult. The ore being mined was more impoverished than at any time since the beginning of the sixteenth century. It was a struggle for survival. But neither the Royal College of Mines in Stockholm nor the master miners in Falun were prepared to give up. Instead, a series of steps were taken to improve the mining and smelting processes. New products were developed.

One of the most significant of these steps, in terms of the future, was taken by Johan Gottlieb Gahn and Anders Polheimer when they took over the operation of the chemical industry at the mine and began to develop it. Most important was a precipitation plant in which the copper in the mine water was precipitated on to iron. But they were eventually also to operate a vitriol boilery, a sulfur plant and a works for the manufacture of red paint.

Ideas for production on these lines originated centuries before. In the case of precipitation, as early as 1626, Andreas Bureus was able to relate that iron was transformed into the purest copper if it was submerged in the mine water. Since then, discussions and preparations had been in progress to adapt the process for larger scale production, which had already been accomplished on the Continent, as in Hungary.

Vitriol was mentioned in the bailiff's accounts in 1540 and vitriol boiling was practiced for at least ten years after that. At the same time, sulfur production was in progress, although undoubtedly on an extremely modest scale.

Swedish manufacture of red paint, a by-product of sulfur production, has been traced back to the end of the sixteenth century when Johan III decreed that the roofs of his castles in Stockholm and Åbo should be painted red.

During the latter half of the seventeenth century, the custom of painting houses red had spread throughout the country. If the financial means were not available to build with bricks, it was possible to gain social prestige by painting buildings red. By the 1680s, red-painted master miners' dwellings had become a feature of Dalarna Province.

The red paint was manufactured somewhat sporadically at the mine from the beginning of the seventeenth century. A hundred years later, operations were organized on a more rational basis.

The bailiff, Anders Lundström, a man of many skills, constructed equipment for a red-paint works and later began production of blue vitriol. Anton von Swab, a member of the Royal College of Mines, and regarded at the time as the country's leading metallurgist in the field of precious metals, was given the responsibility of supervising the new works.

Swab and Lundström were convinced that mine water contained copper vitriol in sufficient quantities to serve as a base for a continuous manufacturing operation. Moreover, there was no reason why the vitriol content should not be used in the production of precipitated copper.

The nature of the chemical process had been understood for several decades. Experts such as Mine Superintendent Anders Swab (Anton's elder half-brother) stated that there was no question of any alchemical transmutation of metal. Instead, the copper particles in the vitriol solution precipitated out and adhered to the iron. Sulfur was extracted by heating pyrites in a retort.

Lundström constructed waterways, channels and sumps, and purchased scrap iron. The works buildings were located at the Creutz Shaft. The red paint plant still stands there today. The little factory also had its own assay chamber and laboratory.

The manufacture of red paint became successful and production increased. Other operations, on the other hand, never grew to any significant extent and operated at a loss. The plants fell into decay.

The change of ownership in 1775 provided a new start. Gahn and his colleagues promised they would organize the operation along new lines and, at the same time, increase its size. For example, more raw material for the precipitation process could be obtained by locating the works down in the mine.

The precipitation was, in fact, the main problem. The mine water showed an increasingly lower copper content and yielded far too little raw material, whereas iron consumption was greater than estimated. Nevertheless, the operation continued for 50 years before it was finally terminated.

It can be claimed that precipitation was the first attempt to produce copper by means of "the wet method", destined to become the method of the future.

Production of red paint became an unqualified success. By the 1760s more than 300 barrels were being produced annually, compared with three or four, ten years earlier, and the increase continued. Falun red was on the way to becoming Sweden's third national color.

The manufacture of vitriol was also eventually turned into a profitable enterprise. Although this was not possible until Gahn introduced new methods. Sulfur production was not an immediate success either. But the method introduced by Gahn after many years of testing subsequently remained in use until the early twentieth century.

All in all, it must be said that the success of the new operation was hardly overwhelming, although it broke even most of the time. This, however, was not the most important factor. The point was that from now on the chemical industry had an indisputable right of abode at the mine. Also born out of this were the crucial preconditions for the mine's continued existence. The byproducts would become Kopparberg's main products.

The scientists

Gahn's initiatives can also be viewed in a wider, national context. This epoch belonged to the visionary entrepreneurs more than any other group in society. When the long period of wars finally came to an end with the signing of the peace treaty in Nystad in 1721, and the nation was freed forever from the burdens of a great power controlling an empire which it lacked the resources to maintain, both in terms of population and economy, recovery was remarkably swift. After a time, Sweden once again became a great power, but in new fields: the natural sciences and technology.

A typical manifestation of the new spirit was the foundation of the Royal Academy of Sciences in 1739. It was here that most of the pillars of the country's intellectual establishment gathered during the eighteenth century. There was great interest in discoveries of a functional nature and "economical and practical matters". The country was in a state of feverish activity. No distinction was made between basic research and its application. All science was at the service of society.

In this spirit, Mårten Triewald was the Academy's foremost authority on experimental physics, Klingenstjerna on mathematics, Linné on natural history, and Wargentin in astronomy. Most had colleagues in their fields who, like themselves, had good reputations in Europe and cultivated regular international contacts. Possibly the most important branch of the Academy's activities was that related to the Swedish economy. Its "Papers" contained numerous articles on practical ways of improving husbandry: information about ploughs, seeding machines, mills, fertilization, seeds and crops. It also covered the art of construction, military technology and shipbuilding as well as discussions on the basic principles of economic policy, on manufacturers and trade issues. And there were items on mining.

The innovators in the latter area were primarily a number of chemists and mineralogists. They came to be regarded as the foremost in Europe in their own fields.

Chemistry and mineralogy have been characterized as the most robust and richly promising sciences in Sweden during the latter part of the eighteenth century. Most of the researchers were connected in one way or another with the Royal Mine Board. Consequently they were all, naturally enough, aware of the practical problems of mining.

This differed from the trend in Europe, where research was oriented to a great extent towards theoretical issues. This was particularly true of gas chemistry, where Black, Cavendish and Priestley managed to produce and document carbon dioxide, hydrogen and oxygen. In the 1770s, with the aid of new techniques, the pioneering researcher Lavoisier was able to show that the generally widespread theory (as was the case in Sweden) which described an indeterminate, weightless substance thought to disappear on combustion—phlogiston—was, in fact, erroneous. This heralded a new epoch for the science of chemistry.

In Sweden, theory took second place. Instead, analyses, documentation and the establishment of systems were given priority. A series of new metals and substances were identified. Due to Sweden's wealth of rare minerals, Swedish scientists of the century were able to discover many more new elements than those of any other nation.

The tradition can be said to have begun with Urban Hiärne, the first chemist in Sweden's history to work independently. As the head of the Royal Laboratorium Chymicum during the 1680s, it was his task to carry out a diversified research program.

A great deal of the work was devoted to the production of medicinal preparations. This was partly for the benefit of the miners in Falun because they were subjected to "many unexpected tribulations, including dizziness, blows, muscle spasms, haemorrhaging of the lungs and similar maladies".

Urban Hiärne's first priority, however, was to investigate the properties of minerals and ores for the benefit of the mining industry. With this in mind, he cultivated the art of assaying with considerable success. This was the method of determining the metal content of a particular type of rock, using a small, specially constructed furnace.

The art of assaying was already well developed by the end of the Middle Ages. In the sixteenth century it had been described by Biringuccio and Agricola. Gustav Vasa showed great interest in the method, primarily in connection with the extraction of silver. When the central mining authority was founded in Stockholm in the 1630s, it lost no time in employing a German assayer. In his work "Archimedes Reformatus", the learned Georg Stiernhielm described how the art of assaying could be employed in connection with the production of coins.

At the Copper Mountain, assaying was practiced at the close of the sixteenth century and, in the 1660s, a proper assay chamber was established at the Mine to determine the copper content of newly discovered ores.

Hiärne did not work exclusively with metals in his laboratory. He also produced "a varnish or balsam, which will preserve wood against rot" and experimented with different types of fuel-saving furnaces. In addition, he set up a project for a vitriol boilery in Falun. The aim was primarily to achieve a better end-product than was usual for his time; such a product could be used for dying clothing.

Hiärne's interest in mining was also motivated by his position as assessor at the Royal Mine Board. One result of this was Sweden's earliest popular manual on mining methods and the art of smelting, embellished with the following ornate and explicit title: "A rather small MINER'S LANTERN, which can be used to enlighten and guide the reader through the dark regions of the mining industry, lit in haste and most humbly presented on the joyous return from Bergslagen district of His Royal Highness, our most merciful King and Lord (1687), by Urban Hiärne".

The subsequent golden age of Swedish chemistry began in 1727 when, after a period of decline, the Royal Mine Board's laboratory was taken over by the knowledgeable doctor and chemist, Georg Brandt. At that time, the laboratory fulfilled perhaps its most important function as a training establishment for young mining engineers. Here, after gaining practical experience in mines and quarries, they received their complementary scientific training from Brandt. The current established method was to gain knowledge of smelting processes in general through laboratory trials.

Brandt was not just a practician. He was also the first of many Swedish research scientists. Perhaps his best known achievement is his identification of the metal cobalt. His scientific contributions also had practical relevance. For example, the production of brass and, most notably, the chemistry of zinc.

Among those who tried their luck as mineralogists was Linné. He had already received his introduction to the art of assaying during a journey to Lappland. It seems feasible to suppose that his interest in mineralogy and, in particular, its practical applications, was his most important reason for visiting Falun, visits which proved to be of such central importance to the rest of his career.

He relates how he "crept among the stones in the mine and at night sat near the smelting furnaces". With Falun as his point of departure, he also embarked upon a special journey to study quarries and mines, "Iter ad fodinas".

On his return to Falun, he summarized the system of classification he had devised for minerals in the treatise "Pluto Suecicus and a description of the kingdom of rocks". At about the same time (1734) he also gave lectures on the subject to a group of young master miners' sons. These were held in the company's assay chamber, and dealt primarily with the art of assaying. Linné also subsequently lectured on mineralogy in Stockholm, in the hall containing the Royal Mine Board's minerals' cabinet. It is said that the audience was so large the floor nearly gave way.

It has been claimed that Linné's attempts at classification were not entirely in vain: they were clear and consistent. In common with his many predecessors, his criteria were the external distinguishing marks of minerals, rock types and fossils, particularly the crystalline structures. His data proved to be inadequate. The resulting system was arbitrary. It was of dubious value when first published in 1735, and definitely outdated when it was included in its final form in his famous synthesis, "Systema naturae", in 1768.

Axel Fredrik Cronstedt's "Approaches to mineralogy" was on the other hand a pioneering work. It was published anonymously in 1758. The main basis of this work was a presentation of the chemical composition of substances. It was necessary, he contended, to isolate the basic elements of each substance.

The work greatly enhanced the understanding of the mineral kingdom. Cronstedt became to mineralogy what Linné was to botany. His work was studied and applied throughout Europe.

It was mainly due to his skill as an analyst that he was able to carry out the task he had set himself. His most important tool in this work was the blow-tube, originally invented in Germany in the seventeenth century. The method involved heating with a flame an extremely small amount of the mineral one wished to analyze, together with fluxes. The chemical composition could then be ascertained from the resulting color changes. The blow-tube became an indispensable tool for chemical analysis.

Cronstedt gained inspiration for his work as a student of Uppsala's first professor of chemistry, Johan Gottschalk Wallerius. Despite his academic position, Wallerius was mainly interested in serving society in a practical sense.

Wallerius was a conservative man. He attempted to reconcile the results of current research with the Old Testament doctrine of the Creation. He made no more notable discoveries but was highly competent in his field and an excellent teacher. He was particularly skilled as an author of handbooks. His extensive "Systema mineralogicum", completed in 1775, contained dated and conventional knowledge, but was comprehensive and organized and even aroused international interest. Prior to this he had already published a work called "Elementa metallurgiae", in which he dealt with mineralogy in general and, in particular, the chemical reactions which occurred in roasting and smelting furnaces.

Swedish chemistry reached the peak of this, its Golden Age, with the work of two researchers, close friends although totally different personalities: Torbern Bergman and Wilhelm Scheele.

As a result of their activities, a rapid stream of new findings in the field of chemistry flowed from Uppsala, particularly during the 1770s. Bergman can be said to have laid the foundations of analytical chemistry. In pursuing this, he carried out innumerable analyses of substances and ores. Among these was the first detailed chemical description of carbon dioxide, which led him to discover a method for the production of artificial mineral water.

Among Bergman's many contributions, his research into the problems of affinity is the most remarkable. He set himself the task of discovering through experimentation how all known substances reacted with each other. Theoretically, it involved as many as 30,000 experiments. His results took the form of extensive tables.

Bergman maintained vigorous contact with leading European chemists, many of whom learnt Swedish in order to study his work. In many cases, their efforts were unnecessary. His work was translated into seven languages, including Portuguese and Russian.

The mineral and rock collection which Bergman housed at his institute was undoubtedly of great value to his students, the practical chemists of the future. The collection provided a rapidly assimilated overall picture of the types of rock to be found at the sites of the country's many mines. Copies of the various chemical-technical aids used at these work sites for extracting metal, the manufacture of sulfur, alum preparation, potash refining and other purposes were displayed in a special cabinet.

Unlike Professor Bergman, Scheele never achieved any social status. When as a famous researcher he was elected a member of the Royal Academy of Sciences in 1775, his formal title was "pharmacie studiosus". He has been described as essentially an empiricist, "the foremost experimental chemist of all time". But, as time went on, his interest in solving theoretical problems became increasingly pronounced.

Scheele is best known for the large number of new substances he discovered and documented. One example is his thesis of 1774 entitled "Manganese or Magnesia, and its properties", described as a distillation of illuminating discoveries. For instance, he documents three new elements: manganese, barium and chlorine. But even more remarkable was his discovery of oxygen. This event took place in the middle of the 1770s. Several other researchers were working along the same lines as Scheele, primarily Priestley and Lavoisier, but it now seems to have been established that Scheele was ahead. The proof of this is found in his correspondence even though the printed account of the discovery, in "Von der Luft und dem Feuer", was not published until 1777.

The period from 1770 to 1775, during which Scheele worked as an assistant in the apothecary's shop in Uppsala, was the happiest of his life. This was due to a great extent to his close and fruitful cooperation with Bergman. The man who brought them together and became a devoted friend of both was the previously mentioned Falun student, Johan Gottlieb Gahn.

Gahn was one of Bergman's very first pupils and, with his practical turn of mind, was no doubt a great help to his professor. When Gahn joined him, Bergman was, of course, a highly respected, versatile researcher, but chemistry had hitherto not been his central interest.

Gahn was also the most important of Bergman's students from a scientific viewpoint. Although, unlike his teacher and Scheele, he never attained international recognition he was not far behind them in terms of knowledge and skill. Gahn made essential contributions but generously left it to his friends to make his discoveries public. He became a member of the Royal Academy without having published a single sentence.

But Gahn played a decisive role in the discovery of the substance manganese. He also discovered baryte (heavy earth), and established that the inorganic constituent of the bones of mammals is phosphoric lime.

Together, the three scientists formed a group that was virtually unique in the history of Swedish science, achieving a series of remarkable discoveries. Their working relationship was so intimate that it is often difficult to distinguish between their individual contributions.

At the beginning of the 1770s, Gahn returned to Falun where after a time he became a master miner. In this capacity, he became increasingly involved in the mining company's operation and general affairs. As a result, he became the most prominent, although by no means the only individual to combine sound theoretical knowledge with the capacity to solve practical problems.

To the same group belonged the previously mentioned Cronstedt who became both technical director and mining superintendent of a large part of central Sweden's mining district.

Also mentioned earlier was Anton von Swab. We have Linné's word that Swab was the most knowledgeable metallurgist in Europe. At the same time Linné maintained that Brandt was the world's foremost "chemicus". A modern specialist, Hugo Olsson, has expressed the opinion that these assessments were well motivated.

Another such figure was the versatile and politically active Daniel

Tilas. His specialty was ore geology. His contributions in this area included analysis of the ores at the Falun Mine. Sven Rinman was a prominent metals analyst whose contributions were mainly in the area of iron. His monumental works, "Approach to the history of iron" and "The Mining Lexicon" are classics. Both were published during the 1780s. The contents of the latter work in particular indicate that he also possessed a thorough knowledge of conditions at the Copper Mountain.

In view of all the theoretical knowledge available and taught at this time, it is natural that those engaged in practical mining at the Mountain were also better educated and perceived their work differently compared with their predecessors. This applied to the highest management—mining superintendents such as Anders Swab, at the beginning of the eighteenth century, and Per Hedenblad, who headed the operation during the latter part of the century, a period during which important initiatives were taken in many areas to modernize the operation. But new ideas often originated from other members of the administration whose greater insight into the operation gave them the opportunity to make independent contributions.

From 1753, there were well-defined educational requirements for active master miners. It was during this year that a mining examination was introduced. Before even being allowed to enter the examination, the candidate had to fulfill certain formal conditions: he had to own a minimum of two mine shares and enough land to support a horse. During the actual examination, the candidate had to demonstrate knowledge of the law and a certain amount of physics and mathematics, but he was examined primarily in mineralogy, chemistry and the art of assaying. He was expected to be familiar with the various types of ore, the method of building a roasting pit, the construction of a smelter, and the building of a smelting furnace. The examination took place in the presence of four experienced master miners chosen by the Mining Court at Falun.

This new attitude applied to all aspects of the operation, not just the operation of the mine but also its organization and economy. Chemistry and mineralogy were seen as important but only if supported by, for example, practical and theoretical mechanics and a knowledge of preparing financial estimates. The government maintained that no aspect of mining could be treated "as a handicraft".

This development was introduced throughout the European minerals industry during the eighteenth century. Advanced tutoring in the mining sciences was offered at establishments of higher education in Germany and Hungary.

Characteristic of the new spirit were the repeated attempts made at the Copper Mountain to achieve a fully comprehensive model of all the deposits and ore veins of the body of ore. Attempts of the same type had been made by Anders Swab and his direct successor Samuel Troili. In 1753, five members of the mine's staff were commissioned to survey the mine's system of ore deposits, which was subsequently meant to form the basis for a systematic mining strategy. These investigations had no immediate results in practical terms. This was also true of the preliminary quarrying plan which Daniel Tilas, undoubtedly the leading expert on the geology of Falun Mine at that time, presented in 1759.

The initiative was of interest mainly as an indication of developments to come. That its success was limited, was to a great extent due to the complex pattern of the body of ore at the Copper Mountain.

The Mine

The impoverished ore meant that even a limited production of untreated copper demanded the processing of huge amounts of material. In addition, the ore had to be sought on new levels which could only be reached at considerable cost. In fact, operations at the Mine were on an even larger scale now than they were during its heyday. Firewood consumption increased and timber felling had to be carried out at increasing distances from the Mine. Although at this time the Royal Mine Board regarded the iron industry as its main concern, it also took a constant and watchful interest in activities at the Copper Mountain.

Technically, the situation at first showed little promise. After the major collapses at the close of the seventeenth century, the mine was described as "a heap of stones".

When Christopher Polhem, already a renowned mechanical engineer, came to the Mountain during the 1690s and eventually became the mine's technical director, he was the object of great expectations. With ingenious inventiveness, he constructed new machines for virtually all applications. Polhem was a master of the techniques of wood construction. In addition to being used for the pumps, the long walking beams, which came into use at the beginning of the seventeenth century, were now used for various types of hoisting devices as well. His new constructions aroused both admiration and amazement in the classic German mines, such as those in the Harz. Polhem also had radical views about the organization of the work. The starting point ought to be a perspective of the operation as a whole; the mine and all its diverse mechanical installations should be regarded as component parts of one large mechanical problem.

However, Polhem's new contraptions had their weaknesses. They were often far too complicated and, as a result, unreliable in operation. The long rods caused jerky operation and involved considerable energy losses. The master miners preferred simpler, more reliable devices. Polhem was undeniably the greatest Swedish mechanical genius of his time, but at the mine his contributions never played a decisive role. His resignation from the Mountain was charged with bitterness on both sides.

When he died, all his remarkable contraptions had already been dismantled.

Most of Polhem's work at Falun was carried out at a time characterized by great suffering and worry at the Copper Mountain. The finances of the realm were in deep disarray as a result of the wars, and the effects also were felt at the Mountain. Coal supplies ran dangerously low. The

mineworkers suffered extreme deprivation and demanded better conditions. The feeling of unity within the mining community was disrupted by serious crises.

The competent and resourceful Anders Swab, who had held the post of mining superintendent since 1714, attacked all these problems with great energy.

From an organizational viewpoint, his main contribution was in connection with the mining procedure. Traditionally, share-owners were let into the mine chambers one at a time and allowed to mine their ore assisted by their own people. In the future, all mining was to take place under the aegis of the company. It hired mining hands, organized the hoisting devices, and provided tools and firewood for which the share owners were obliged to pay a fixed rate. In time, measures were also taken to ensure that all share-owners would be allotted ore of approximately equal quality. Consequently, there was no longer room for the old individualism. When the reform had been fully implemented in the 1740s, after the Crown's share had been set aside, all that remained was to share the rest of the ore mined between the share-owners on the principle of "fair shares for all". The mine had become a single operating unit.

Since the Polhem era, technical development had progressed more slowly. In many areas, such as hoisting devices for both water and ore, the old tried-and-tested methods were adhered to.

However, to meet the demands created by the need for a more large-scale operation, a new, deep shaft was required. Excavation work was begun in 1716. It was located on the western edge of the Great Pit and represented the largest single initiative to be taken in one hundred years at the Copper Mountain. It was not completed until 35 years later, but after 20 years, the bottom originally estimated was reached and the rate of hoisting could be speeded up. When the construction of the pumping machinery, the Great Pump, capable of raising water from a depth of 220 meters, was finally completed, there was a considerable sense of relief. The machine was driven by two walking beams with an overall length of

360 meters. Vertically, it operated with 25 sections resting on timber bases.

The shaft was named after King Fredrik. It quickly became the main channel of transport at the mine. The inauguration, patriotically held to coincide with the birthday of the new king, Adolf Fredrik, was opened by a 48-gun salute, as the water was released onto the drivewheel at the dam and the pumps began to work. The entire construction staff, 101 persons, were invited to dinner afterwards. The town's bedridden poor and patients in the mine hospital were also remembered.

The celebration was justified. At the beginning of the 1750s, the ore deposits were thought to be ample. The amount raised was exceptionally high. It amounted to 200,000 barrels annually.

Yet the capacity of each individual machine was not particularly impressive. In the deep hoist shafts, they reached a maximum of 10 barrels per hour by the middle of the century.

Alongside the large water-powered machines, the horse-drawn winches continued to play an essential role in hoisting operations throughout the eighteenth century. As the century wore on, the Sala example was followed and horse-drawn winches were also introduced down in the mine. By the middle of the 1770s, there were 17 horses working underground. Travellers thought the horses were content down the mine. "Their loud whinnying showed that they had become accustomed and thus totally reconciled to their dark abode."

For removing water, apart from the installation at Adolf Fredrik Shaft, there was an older machine in the so-called Wrede Shaft. Nevertheless, the two pumps combined could barely keep the deep mine from flooding. There were many complaints about the old Wrede pump. A particularly serious danger was that the mouth of the pit lay open and totally unprotected. Therefore, there was a constant risk of icing up in the winter. During periods of severe cold, the pumps had to be operated continuously to prevent them from freezing. If, despite this measure, they froze and stopped, the ice formed rapidly along the entire length of the machine so that pumps, pump seatings, steps and beams became one unmanageable mass of ice which had to be hacked away or melted with hot water.

Falun's master miners may appear conservative because they did not attempt to install a steam-driven pump. This was done by Mårten Triewald in Dannemora, although with only very limited success, as early as the 1720s. The reason is clear. In contrast to the situation at Dannemora, in Falun, most of the time there was sufficient water-power from the lake system above the mine which had been regulated for centuries.

Fires in the mine

During the seventeenth century, there were no timber constructions of any size erected in the mine. A determining factor was the risk of fire. "Fire setting" and shaft timbering did not go well together.

After the major collapses of the 1680s, a new situation arose. In spite of the risks, timbering work had to be carried out amongst the rubble from the collapses. Severe punishments were introduced for carelessness. Around 1700, anyone who negligently ran with a lighted torch in a timbered pit was sentenced to run the gauntlet six times, even if no damage resulted from his action.

In the early eighteenth century, comprehensive timbering was installed to support the walls of the mine which were on the point of collapsing. Extensive communication galleries were built in the debris. This was a skilled task. The technique was imported from Germany. From the 1730s onwards, the mine was filled with large chambers and galleries lined with timber. One reason for the abundant construction work was that the engineers thought they had learned how to protect the timbers with covering walls.

In 1760, however, the old misgivings were justified. Fire broke out in the timbers of one of the shafts. The fire raged for many months and threatened both the buildings of the Mine administration and the town. This was the start of a series of catastrophic fires in the area. During the same year Stora's miners' lodge was reduced to ashes. The following year, the town of Falun suffered the worst fire in its history.

From this point on, as far as the mine was concerned, fire replaced collapse as the major risk factor. In 1768, there was another fire, this time more extensive and of longer duration than before. It started in a section of the mine containing many wooden buildings and took six years to put out. During this period, many working rooms were inaccessible. Timbering was destroyed, columns and fixings collapsed, gangways were blocked.

However, the worst disaster, was yet to come. This was a fire which broke out in 1799 and was not put out until the early 1820s.

Mining

But there were also brighter aspects. The efficiency of actual mining methods was improved in many respects.

The improvements were to a large degree the result of the many study trips Swedish master miners had made to the Continent since the 1690s. Because of these journeys, Swedish mining experts were well informed about the methods used in the famous continental mining regions in the Harz mountains, Mansfeld, Saxony, Bohemia and Hungary. In addition, a lot of information could be gained from the strongly German-influenced operation of the copper and silver plants in Röros and Kongsberg in Norway. At the Copper Mountain, the actual body of ore, the large aggregate of pyrites, was so formed that, ever since mining operations had first moved underground, the natural method of working had been to fire and excavate high, wide chambers, supported by pillars which were left behind as the work progressed. This was criticised as primitive and inefficient. However, the mining industry's travellers were able to report that, contrary to some opinions, the situation at the Copper Mountain was by no means unique. Mineral deposits on the Continent with similar potential, such as the tin mines in Bohemia and Saxony, were mined using similar methods. For a long time they were regarded as the only ones viable.

In the Falun Mine, the low ore content of the rock resulted in the introduction of manual sorting. In itself a simple measure, it involved separating barren rock from ore-bearing rock at an early stage. Prior to 1700, this had only been carried out in exceptional cases. Initially, it was intended that dressing should be carried out down in the working rooms. The unproductive rock should be left in the mine. Subsequently, special sorting staff were employed at the winches. The work was quite costly. It was difficult to find a really satisfactory solution.

The new mining methods, first introduced during the latter half of the eighteenth century, were much more radical. They included various forms of stope mining and, especially, overhead stoping (Fürstenbau). In both cases, the advantage lay in the fact that all available ore could be extracted from the mine.

The first time the cut-and-fill method was mentioned in connection with Falun Mine was 1753. During the 1760s further advances were made. But it was not until 20 years later that the new methods achieved a real breakthrough. Gunpowder blasting was used for virtually all stope and overhead-stope mining. This was another situation where a method of working which was well known at the Mountain was only adopted after lengthy consideration.

Apart from stoping and overhead-stoping, "blasting" was the greatest technical innovation. The first breakthrough took place during the 1710s.

The only individual at the Mountain who was totally comfortable with this new method was Mine Foreman Hans Öhlbom. For a fee of 5 silver dalers for each shot, he carried out the blasting now being permitted in nearly all parts of the mine. His biggest task was a further deepening of the Fleming Shaft.

Several setbacks occurred when the Royal College of Mines introduced a complete ban on blasting. However, by the 1730s, it was without question an established technique, although great care was taken and only shallow bore holes were permitted. Blasting methods advanced along the lines of those employed on the Continent. Wide holes, packed with comparatively large quantities of powder, were replaced by narrower, deeper holes with smaller charges. Increasingly effective tools were used for drilling. When purchasing powder, preference was given to local manufacturers. Among the inhabitants of Husby, in southern Dalarna, were some competent powder manufacturers whose product was effective and, in the opinion of the Copper Mountain's master miners, superior to that of other manufacturers. Kloster, a local powder works, was also one of the suppliers.

The old "fire-setting" methods were still obligatory in the larger chambers and continued in the old manner alongside the powder blasting.

Due to the expansion of the main mine, which contained more than 100 chambers, a greater number than ever before, it became necessary during the eighteenth century to limit fire-setting to certain 'fume-days'. When the fires were burning, large sections of the mine were inaccessible. The comparatively large numbers of horses which now worked down in the mine had to be raised with great difficulty to the surface. Considerable efforts were made to organize some rooms which could serve as a refuge for both people and animals when the fires were burning.

These efforts to improve production were rewarded with some, if limited, success. By the middle of the 1770s, production was rising, to reach a peak approximately 10 years later. From then on, it stayed at well above 900 tons annually for an extended period. This was the best result recorded since the period prior to 1720. The struggle for survival had proved successful, for the time being at least.

The smelting operation

The craftsmen responsible for roasting and smelting the ore at the Coppar Mountain have often been described as conservative and obstinate. They continued to work at their trade, employing traditional methods, at least until the 1770s. This period was not totally uneventful, however.

The years leading up to and beyond 1750 showed some improvements in the process. The innovators were two furnacemen, Jöran Blix and Hans Kies. Their improvements were a direct result of practical experience and included better handgrips and some minor modifications to the furnaces. Their results were, nevertheless, worthy of note. They led to an increase in smelting capacity. More than half as much material again could pass through the furnace every 24 hours. The news was received with general satisfaction. The master miners declared that Blix and Kies' ingenuity had improved the smelting process to the extent that further improvements were impossible to imagine.

This opinion was not generally accepted.

Exactly midway through the century, Axel Cronstedt carried out an ambitious survey of conditions in the precious minerals industry and at the same time proposed measures for improvement. One result of his work was that the company decided to award silver goblets each year to two particularly deserving furnacemen, which would have the dual effect of raising the status of the craft and encouraging employees to make suggestions for improvements.

That much remained to be done was shown very clearly by a new, extremely thorough investigation of the smelting process at the Copper Mountain which was carried out by Samuel Gustaf Hermelin in the summer of 1770, at the request of the Royal College of Mines.

The report he subsequently submitted to the Royal College was harshly critical and caused great uproar among those responsible. The immediate consequence was a sharply worded note to the Mine Court in Falun. At the Royal College of Mines there was concern about "how obstinately, negligently and indecently the Copper Mountain people run their smelting operations". Their "vanity and self-indulgence" had to be checked.

The origin of Hermelin's conclusions lay, of course, in the fact that substantially more was now known about the preconditions for the smelting process than just a few decades earlier. Hermelin's proposals for improvements included using the metallurgical precepts of the time. These were based primarily on an analysis of the chemical reactions of the smelting process.

Hermelin realised that his investigations would not lead to any practical results worth mentioning if they were not followed up continuously with more work of the same nature. As he was unable to continue the work himself, he sought a successor. At Christmas, 1770, he was in Falun, where he met Gottlieb Gahn who had travelled up from Uppsala during the vacation to visit his parents. This meeting was to change Gahn's life and point him in a new direction. As related earlier, at this time the 25-year-old Gahn was firmly anchored in academic circles in Uppsala and was looking forward to a promising career as a researcher. It therefore required a great deal of persuasion from Hermelin before Gahn could be prevailed upon to accept the task. Once this was achieved, however, his commitment was whole-hearted. A new phase of the work to improve the smelting operations in Falun, founded on scientific principles, was initiated in the early years of the 1770s. Gahn was the most prominent chemist to date to become involved with the smelting process at the Copper Mountain. As such, he was an early example of the academic researcher with the capacity to convert his knowledge to practical use. To a great extent, his work took the form of laboratory research. After only a few months he established a small experimental workshop in a forge at the mine. His ambition was to clarify the entire theory of the copper-producing process, thereby preparing the ground for the practical improvements he wanted to propose. He never got that far, but the changes in the process which took place as a result of his work were important enough in themselves.

As a consequence of his research, furnaces were constructed taller than previously. Tall furnaces became commonplace at the Mountain around 1774. According to Gahn, such furnaces provided greater heat, easier smelting and, consequently, savings in coal. In addition, he improved the composition of the flux in the furnaces by adding more lime. Logically enough, together with Hermelin, he also tackled the problems connected with refining at Avesta. Gahn was a strong supporter of the idea that the entire operation at Avesta should be relocated to Falun. This did not take place, however, until well into the nineteenth century.

However significant Gahn's contributions may have been, it must be said that they did not involve any revolutionary changes. The industry continued to retain its traditional basic structure and there were still foreign observers who felt the operation lacked rational organization.

Around 1775 Gahn's assignment could be regarded as complete and by then he had already turned his attention to other tasks. As we have seen, these were mainly in connection with the chemical industry. However, he did not forsake the theoretical problems of smelting. Later in the 1770s, he was successful in his attempts to improve blasting methods in the smelting works.

Parallel to his versatile private activities, Gahn devoted considerable time to acting as an advisor in nearly all matters relating to the company. His mature and well-balanced judgements and, most of all, his uniquely diverse insights and wealth of ingenuity, made him virtually indispensable to his peers at the Mountain.

He maintained contact with more centrally located researchers, to the mutual benefit of all. In his later years, he became a fatherly friend and mentor to the well-known chemist J.J. Berzelius, particularly with regard to the blow-tube technique of which Gahn was a master.

Gahn did most of his work in his extensive and unusually well-equipped laboratory, which boasted its own well-stocked library. Given these facilities, it was natural that the first general technical college for mining science should be located in Falun. This establishment was founded at the beginning of the 1820s with one of Berzelius' most prominent pupils as the first principal. At the School of Mines, regular analyses of the mine's ores and the composition of the products of the smelting works were carried out. The fact that the School inherited Gahn's library and collection of instruments proved of priceless value. The tradition of research thus established in Falun has since expanded, especially in terms of Stora's own operations, and now focuses on new areas not related to mining operations.

Trend of ownership

Towards the close of the seventeenth century, there were almost 800 share-owners at the Copper Mountain. Approximately two-thirds of these were responsible for the total copper production—'active master miners'.

A hundred years later, the number of share-owners had almost halved, while the percentage of active master miners had, proportionally, decreased much more. They now accounted for less than one-sixth.

Behind the figures lies a development of central significance. In the seventeenth century, there was still an organization at the mine which was a remnant of the Middle Ages. The miners formed a sort of guild. The mining operation was still a corporate company in peasant hands. Participation involved more than the mere ownership of shares, it also carried an obligation to participate in production. The way in which this structure disappeared to be replaced by a new one, which dictated that the master miners were no longer allowed access to the mine, has been described already. Ownership of shares had acquired a new significance. Towards the end of the eighteenth century, there was an increasingly pronounced trend towards its becoming an investment object. The shares in the mine were owned by wealthy people all over the country. Linné is one example. The actual smelting operation was handled by a small minority whose output per capita had doubled compared with a hundred years previously, and amounted to more than eight tons of crude copper annually.

Around 1800, the mine found itself in a position which increasingly resembled the situation in the Middle Ages, when the aristocratic association owned the majority of the shares in the mine without being directly involved in its operation.

Complaints about the low copper-content of the ores reverberated throughout the entire eighteenth century. It was said at the beginning of this period that, in the old days, there was cause to drink Spanish wine from wooden mugs. Now, equal quantities of red and salty tears should be shed over the harshness of the times.

It is indisputable that, during the course of the century, ore was mined which was not worth smelting. Some foreign observers were surprised that such a meager deposit as Falun should continue to be worked at all. There were many examples of continental copper plants which benefitted from substantially richer ore. Such works existed, for example, in Hungary and in the Harz. But those plants, which supported the greater part of the German copper industry (for example, Riegelsdorf, Thalitter, Mansfeld), were actually not much richer than those in Falun.

In Sweden there was an awareness of the dangers of going too far. The operations should not be driven beyond "their natural boundaries".

With such conditions, it was no surprise that the value of mine shares declined.

During the 1770s, a share was worth approximately 1,200 copper dalers. This was the same value as 50 years earlier. At the same time, money had declined in value by 50 percent. In relation to the return, however, the price was not too high. Shares in the Mountain were always said to possess a true capital value.

The Crown's profit on the company was considerably higher than that of the shareholders, however. On average, the shareholders received a sum which swung between a fifth and a half of the Crown's income. The total sum distributed was usually in the range 700,000 to 800,000 copper dalers annually.

It must be added here that the operation at the Copper Mountain could count on support from the Crown in several different forms. Firewood and coal was supplied to the Mountain from 13 parishes at an extremely favorable price set by the State. In addition, until 1770, the Crown provided a large team of officials to handle the mine's administration. Apart from senior management, there were the ordinary officials, about 40 persons, ranging from works inspector to pump guard. Almost as many, about 30 at any rate, were also employed as deputies on an 'ad hoc' basis. In 1770, these Crown officials came under the authority of the company. The increase in costs for the local authorities was compensated to a considerable extent by reduced government taxes.

The copper market

During the early years of the eighteenth century, nearly 800 tons of refined copper was produced annually at Avesta from Copper Mountain ore. A further 50 to 100 tons were produced at other copper works. From the beginning of the period up to 1760, on average, not less than half the output was used to mint coins and minting continued to be significant for a further ten years. The bulk of the remainder of the copper went to the brass works.

The amount exported rose during the period from approximately a quarter to reach about 60 percent of total production in the 1770s. By far the most important product throughout the years was brass wire. The quantity was, as a rule, twice that for the other two commodities, i.e. the pure refined copper and the minted metal combined. In accordance with the economic thinking of the time, the brass operation, which in fact involved production of a refined commodity, received considerable support, despite the fact that the process required a significant import of galmey, due to the failure of attempts to build up domestic production. The major importer was France.

Copper and brass prices were high at the beginning of the period and, after a shorter period of decline, remained satisfactory into the 1760s. During the course of the century, an increasing number of foreign competitors appeared, but from the Swedish point of view, prices and demand continued to be favorable until 1770, when a long-term decline started. As mentioned previously, the company was granted tax benefits which, combined with improved efficiency, resulted in increased production.

The original practice of using copper sheet to produce coins continued. Despite their impractical shape, they played an even more significant role than ever during the first part of the eighteenth century. They were not merely necessary for regulating prices, but also, as had been the case during the previous century, were required to fulfill the needs of money circulation. The number of coins minted was large enough to allow a certain amount of export as well.

Coins were not only minted from newly refined metal. Old roofs from many churches also yielded raw material. At the end of the period, vessels for distilling liquor also provided raw material, as they had become worthless after the ban on home distillation of spirits in 1772.

The last year to be stamped on any coinage minted from copper sheet was 1768. But minting continued using old dies. In 1777, the metal coins ceased to be legal tender. The minting which took place subsequently was intended exclusively for export.

Towards the end of the eighteenth century another type of production was started at the Copper Mountain. This time it was a silver plant. The silver content of Falun ore had been well known for a long time. Discussions concerning its extraction had been in progress for at least 200 years. The method usually used to extract silver from copper, segregation, had been used on the Continent since the middle of the fifteenth century. In Sweden, sporadic attempts were made along the same lines without much success, by Hermelin, among others, in connection with his analysis of the copper process.

The new plant, which started operating in 1789, used a totally different method based on the availability of galenite at the mine. The ore's gold content had also been noted at an early date and, for the time being, led to the extraction of modest amounts of gold at the silver plant.

Taken as a whole, the measures adopted during the latter half of the eighteenth century produced favorable results. The mine was able to survive with a reasonable level of profitability. At the same time, it was clear that it had reached, and passed, its peak. The master miners would have to look elsewhere for future income. Almost every kind of industry in the region had for hundreds of years been oriented towards the Copper Mountain. People journeyed there to seek work, they sold their firewood there, they went there with their wares. When all was said and done, the master miners' essential operating base was not the Mountain with its limited resources, it was the entire region with its forests, mountains and running water.

1897

The optimists

ON THE MORNING of May 15, 1897, the people of Stockholm awoke to an overcast sky. However, the clouds gradually broke and by the afternoon the city was basked in sun. This was an important omen, since this was the day on which the most remarkable event of the year—the inauguration of the great Stockholm Art and Industry Exhibition—took place.

The opening of the Exhibition was marked by a magnificent speech from King Oscar II, who in the same year celebrated his silver jubilee with his two nations, Norway and Sweden. The Royal address was followed by a short cantata composed by the 26 year-old Wilhelm Stenhammar, later to become a composer of international renown. The words were by Carl Snoilsky, a famous poet and a member of the Swedish Royal Academy. This was the era of world exhibitions; virtually every metropolis had been the venue for a wide-ranging display of the achievements of various countries in technology, handicrafts and art.

At the London Exhibition of 1851, the first of the series, Sweden had been represented mainly by handicrafts. But now the nation felt it could compete with others in every sphere.

The country had assigned its best resources to the task. At the final stage of preparations—after discussions lasting 17 years—the organization consisted of 18 national committees, plus 28 regional committees, which covered the entire country. Ultimate authority lay with the organizational committee which was chaired by the Swedish Crown Prince. The industrial section was divided into 180 categories, representing a general pattern of the country's enterprises ranging from handicrafts, mining and iron goods, forestry products, scientific instruments and engineering products to the youngest industry—the generation and transmission of electricity.

Prince Eugene, who in his own right was an accomplished artist, took charge of the art section, which broke all contemporary records in terms of range and foreign participation.

Located on the outskirts of the fashionable Royal Deer Park, and consisting of more than 100 pavilions, the Exhibition encompassed more than 200,000 square meters, equivalent in area to a small Swedish town of the period. The largest of the pavilions, an industrial hall with a cupola and four minarets, was designed by the 'imaginative and highly innovative' architect Ferdinand Boberg, who was also responsible for a number of other works.

Down by the water, the organizers had built a reconstruction of the central section of the Old Town of Stockholm, as it might have appeared towards the close of the sixteenth century. The unfinished Nordic Museum gained a highly praised temporary wooden extension, designed by the architect Agi Lindegren, in which the institutes of higher education presented their programs.

A number of larger industrial companies had their own pavilions: these included Separator, Sandviken, de Laval, Finspång and Bofors. However, contemporary descriptions confirm that the foremost site was Stora Kopparberg's pavilion 'not least because this company represents our country's oldest and largest company but also because of its size and sophisticated equipment.'

The building provided further artistic testimony of the originality and creative talent of Ferdinand Boberg.

The central, circular section of the house was made of wood, with columns made of of thick trunks and was thus in itself an exhibit of the company's most important product group. The ceiling and part of the walls were covered with wooden tiles painted in blue and white; the base was executed in 'a reddish granite imitation'. The exterior was painted gold, red and white.

The interior was marked by what was the Company's other important area of production: iron and iron products. Between the bundles of iron rods lay rows of ploughs, anvils and other tools.The walls were adorned with rosettes of horseshoe-nails, axes, pickaxes and sledgehammers. The windows were encircled by axes, and the chandeliers were formed from hammers and chains. In the middle of the ceiling hung a colossal white wooden ball decoratedwith horseshoe-nails and studded with picks, giving it the appearance of a gigantic burdock.

Among the many exhibits, special mention was made of a drinking horn which was more than a meter long and had been cast in one piece from Domnarvet steel plate. The drinking horn is now on display in the Company's museum at the Falun Mine. The same applies to several highly acclaimed objects: King Gustav III's mining suit, models of Polhem's hoists and haulage machines, a glass model of the Falun Mine in cross section, and a flake of gold.

Boberg's 'blue cave' also gained much acclaim. This was a romantic representation of a mine inhabited by 'tomtar' (gnome-like creatures from Swedish mythology), evocative of the era.

Carl Sahlin, the Company's exhibition manager, and his zealous assistant, Ludvig Zethelius, had spared no effort.

The exhibition attracted large crowds, 'exceeding the wildest expectations' (King Oscar II). When it closed after five sunny months, it had been visited by more than one million people. Of these, some 11,000 visitors had travelled at a reduced price to Stockholm in 20 special 'popular trains' from various parts of the country. King Oscar was himself visited by more than 27 royal personages, two of whom were kings. This year, he noted, had been the happiest one of his life.

The enthusiasm for the event was widespread. According to the Nordisk Revy (a periodical) it aroused 'joyous feelings of national pride'. "This year we Swedes became conscious of our own value, our own ability, and our own resources," wrote Bishop Rundgren. Ellen Key, a well-known authoress and cultural critic, was sure that even the masses must have comprehended 'the incredible combinations of national strength and intelligence that were represented in the large industrial and inventors sections'.

"Nothing else can better demonstrate the power of the upswing noted by Swedish industry and commerce in recent decades", wrote the art- and theater critic, Edvard Alkman. The conceptual approach is described in great detail in the introduction to A. Hesselgren's extensive description. The author feels that the history of mankind is like "a continuous chain of discoveries and progress", although this has never been so clear as during the last century which, continues the author, "has been characterized by substantial and wonderful progress in all areas of culture". The century could be characterized as 'a golden age of constructive industry'.

Not least, Sweden could "show the proud results of its successful efforts towards constructive labor, a manifold increase in population, in national wealth and resources".

The author's unconstrained enthusiasm was justified.

Within just a few decades communications by land and sea had been revolutionized. Man was now also capable of travelling by air. The combustion engine had come into practical use. The great automobile race in 1894 between Paris and Rouen had marked a breakthrough. The electric bulb had been in existence since the 1870s; electricity plants since the 1880s and now 3-phase alternating current and high tension transformers were in use. A power station had been built on the Niagara. Steam turbines and diesel engines were in operation. Marconi had established a company to develop wireless telegraphy. The new iron manufacturing processes, with such names as Bessemer, Martin and

Thomas, had provided the conditions for the manufacture of cheap steel on a much greater scale and for many new applications: rolled steel girders, seamless tubes, skyscrapers, and even an Eiffel Tower. Agriculture was also beginning to be mechanized, with the introduction of threshing machines and tractors.

In industry, it was now possible to produce sulfuric acid by means of the contact method, artificial silk from viscous pulp, aluminum through electrolysis, and oil could be more effectively exploited by means of cracking. The first gramophone record was made in 1887, and the following year marked the advent of the box camera and celluloid film.

The author's description of the Exhibition was also correct in stating that Sweden had taken a place of honor in this comprehensive picture of peaceful efforts. Achieving greater success than in many other countries, the people of Sweden had benefited from an economic boom which was to prove unusually beneficial and protracted.

During the latter half of the nineteenth century, the natural sciences in Sweden experienced unprecedented growth in academic posts. The Royal Technical Institute in Stockholm had been extended and received a new charter emphasizing its scientific character. New subjects within mining engineering had been established, in addition to electronics and electrochemistry. Chalmers Technical College in Gothenburg was also extended.

The favorable research climate in the country resulted in a series of remarkable efforts by such men as Svante Arrhenius within the new field of physical chemistry, Robert Thalén, a pioneer in the research of terrestrial magnetism, Anders Jonas Ångström, an innovator in the field of spectrography and, not least, Peter Klason, whose achievements included a description of the central chemical process in sulfite cooking.

Many other feats could also be mentioned, those of such as the renowned geographers and discoverers. The year 1897 was also the date of the Andrée polar expedition.

In addition to those noted above, one could also add the names of many inventors, foremost among these being Alfred Nobel (dynamite), L.M. Ericsson (telephony), Gustaf de Laval (separators), Jonas Wenström (alternating current) and C.E. Johansson (precision measurement).

La Suède

More cause for rejoicing could be found in the ambitious and comprehensive profile of Sweden which was presented in the 1,000-page publication entitled La Suède, edited by the well-known statitician Gustav Sundbärg and completed before the great Paris Exhibition planned for 1900. The book was also published in Swedish and English.

Here one could learn how a whole new series of industries had been established within the preceding ten years, while older industries had increased their operations many times over. The railroad network, the largest in Europe in relation to the size of the population, had solved the problem of the great distances in the country and had opened the interior to international trade. In the meantime the products supplied were being processed to an ever-increasing degree. The general well-being was described as constantly on the increase.

This claim could be supported by statistics. While less than 20 percent of the population derived their living from industry and trade in 1870, this figure had almost doubled by the end of the century. Thanks to the 'explosive' development of industry and commerce, the value of all buildings in the country had more than doubled between 1862 and 1900, while income from capital and labor had increased by more than 350 percent.

In a review of the various industries, the authors noted that the production of ironbars had increased from 213,000 tons in 1880 to 336,000 tons in 1899. Progress within the mechanical industry was outstanding. The output of metal goods, machinery and instruments increased seven-fold from the beginning of 1870 to the end of the 1890s.

Sawmills and groundwood mills, with an output valued at SEK 140 million, accounted for a larger amount than any other branch of industry. Calculated in cubic meters, exports had doubled between 1870 and 1899. Sales revenues noted an even more favorable trend. Wood pulp, still insignificant in terms of production volume in 1870, and worth less than SEK 2 million by 1880, had grown into a major industry by 1890, with output valued at almost SEK 21 million.

"As regards the quality of chemical pulp or cellulose, Sweden is most likely the foremost country in the world", the author confirmed.

Developments within paper manufacturing were hardly less dramatic. Manufacturing volume at the end of the 1890s was more than five times the figure thirty years before.

The recovery of energy from the country's water resources appeared to be a promising future industry particularly since "the water's energy, when transformed into electrical power, permits itself to be transmitted over quite substantial distances—some tens of kilometers and more".

In his description of Sweden as a whole, the author—at least in the French edition—rises to almost poetic heights. "The means of communication," he assures his readers, "have shortened the distance between even remote villages and have permitted close connections between the country as a whole and, via underwater cables, with the outer world". Telephone and telegraph cables spanned the country with their lines. The railroads joined the grain fields of Skåne in the South with the mountains of Lappland in the far North, the Baltic coast with that of the Atlantic, while high-speed steamers maintained regular connections with neighboring countries. "The increasingly sophisticated lighting system (using gas and electricity) has assisted in counteracting the depressing effect that the darkness of winter has on work".

The poet and literary critic Oscar Levertin, who is hardly renowned as an admirer of industry, wrote in a review that "modern life, which can easily appear prosaic when viewed in detail, aquires true greatness when one, as in this book, gains an impression of the whole. A statistical work, filled with names, figures and tables, is least capable of leaving room for lyric poetry, but nevertheless, I feel that this book carries the sound of a large choir and orchestra playing a hymn to the future of Sweden, a hymn that fills the senses with devotion".

At the Royal Deer Park this surging music of the future was represented in tangible form by the many buildings and exhibits.

Stora Kopparberg

Among the numerous companies described and praised in the various publications issued at the time of the Stockholm Exhibition, Stora Kopparberg, as noted above, consistently received top honors.

In La Suède, the Company was characterized as "one of the most interesting phenomena in Swedish economic history". The author notes that the company in Skutskär had built up the world's largest sawmill and that Domnarvet's steelworks were by far the largest of their type in Northern Europe. The Company's assets were estimated at about SEK 40 million. He drew particular attention to the fact that the Company since 1895 had increased the dividend from 8 to 14 percent over a period of five consecutive years: "moreover, the Company has established several new factories by using its own earnings rather than by borrowing the necessary amount".

In the official account of the Exhibition, the iron expert—and later Professor—Erik Odelstierna eulogized Swedish iron. He maintained that a gigantic step, comparable with the transition from the Bronze age to the Iron age, had been taken in Sweden since the 1860s. Thanks to the capacity and success achieved in processing, it was now possible for Swedish manufacturers to produce the world's finest crude iron and bar iron, raw material for "an excellent, completely fault-free product".

The conditions of workers had improved dramatically. "There was a time", says Odelstierna, "not long before 1866, when smiths at many ironworks had to steal iron and sell it quietly to the farmers so as not to starve, while nowadays there is hardly a blacksmith's apprentice who does not buy a bicycle."

Odelstierna believed that Stora Kopparberg's manufacturing operations which were "on as large a scale as those of foreign companies" made the Swedish master miners "equal to the iron lords of England, Germany and America, in the shadow of Domnarvet's name". Many of the high-quality samples exhibited, he said, "proved a source of wonder to the admiring visitor".

When the Exhibition closed, Stora Kopparberg was presented with five gold medals. The Company was praised for "the great development of iron and steel production, for the implementation of the basic Bessemer process in Sweden, for the products exhibited and for its improved carburization process, for excellent red paint, blue vitriol, ferric sulfate, sulfur and sulfuric acid. As well as for its excellent model for a carbonization furnace, and for its charcoal samples, and planed and sawn wood products".

The Thomas phosphate won a silver medal, while 'good sulfate pulp' received no more than a bronze.

The Company's central position within many important and expanding sections of Swedish industry could hardly have been better demonstrated: the Company's management could feel satisfied.

Background

The development noted by the Company during the nineteenth century was nothing less than revolutionary. The operations which were important in 1897 had hardly been a part of the Company's program a century before, or else had fulfilled some other role.

It is often mentioned that Stora Kopparberg made its first forest acquisition as early as 1724 and set up an ironworks in 1735. The forging of tools was also conducted on a modest level as early as the 1640s. All these cases, however, involved a limited initiative and cannot be regarded as a conscious effort designed for further expansion.

The Company continued to cover its wood requirements primarily through leasing forest land. The Svartnäs ironworks, established to meet the Company's own requirements with respect to tools and other goods, proved to be unprofitable, and were leased out for most the eighteenth century.

A turn-around occurred during the initial years of the nineteenth century with the gradual repeal of chartered privileges, at the same time as the market for iron, at least for the time being (i.e. the beginning of the century) appeared favorable, not least because of the American market. Demand at this time was for conventional German-forged bar iron.

At this stage almost the entire output of the Company's iron was exported, as was also the case for the Swedish iron industry in general. The same conditions applied throughout the whole century, albeit on a slightly diminishing scale. Sales were made through trading houses on behalf of the Company up to the end of the 1870s, after which it began to sell directly to foreign customers.

Even before 1810 the Company had begun to expand at Svartnäs and had applied for permission to set up four ironworks.

When Domnarvet ironworks were completed in 1878, the Company had already been operating blast furnaces at six locations during shorter or longer periods . Two of them were still in operation. At the same time, the Company had also produced bar iron at nine locations, three of which continued to operate for a few more years.

The outlook for the country's iron production deteriorated during the earlier decades of the century. But although foreign competitors were capable of producing cheaper iron in larger quantities due to the use of fossil fuel, Swedish ironworks enjoyed a competitive advantage in terms of quality. Consequently, they continued to manufacture in accordance with the traditional method in many small separate units. The latter was unavoidable, since each ironworks required plentiful supplies of charcoal. This frequently resulted in unreasonably long transportation distances. In the mining district this could entail distances of 150 kilometers before the ore reached a loading station in the form of the finished product.

Stora Kopparberg's acquisition of ironworks proceeded without any clearly defined plan: instead, it was most frequently a question of opportunism. Naturally, the units expanded, but from the international viewpoint they remained insignificant. At the Company's smelting works the average production during the first quarter of the eighteenth century was no more than 340 tons. By means of hot blasters and, primarily, more blasting days, output doubled towards the mid-nineteenth century. During the period from 1851 to 1875 production reached 1,250 tons. These figures surpassed the national average.

The yearly production of bar iron was about 200 tons per ironworks at the beginning of the century and had not increased much by the 1850s. After 1850, production continued on a larger scale at the three or four remaining ironworks.

In general, Stora Kopparberg's status among Swedish iron producers was rather modest. With an annual average production of 3,400 tons of pig iron, the Company accounted for less than 3 percent of the country's production in the earlier part of the nineteenth century. Despite the fact that from 1851 to 1875 the Company reached about 5,700 tons per year, the percentage share had declined. The production trend for bar iron was approximately similar.

Nevertheless, Stora Kopparberg was one of the country's largest manufacturers. In 1844, it ranked number two with a charter-based production of about 2,000 tons of forged iron, produced at eight ironworks, whereas Uddeholm, with operations based on twelve works, produced almost double the amount. Carl de Geer, with six ironworks in the province of Uppland, and S. von Stockenström, with the Fagersta ironworks, were approximately equal in third and fourth place,

Perhaps the Company's most remarkable initiative during the nineteenth century, prior to Domnarvet, was Lindesnäs.

These ironworks were acquired in the early 1850s and were a focal point for an expansion of iron production during the subsequent years. Investments made in the works included a Lancashire forge and a rolling mill for bar-iron, the second of its type in the country.

This mill was to form the core of what was referred to as the Company's Western ironworks. 'The Eastern ironworks' consisted of Svartnäs, a blast furnace established in the 1820s, and a Lancashire forge, dating from around 1840. All these units were regarded as offering good potential for survival, with sufficient wood resources.

But the climate changed. At the beginning of the 1860's Company Management was extremely doubtful about future developments. This decade saw the widespread closure of Swedish ironworks. In response to this tendency, Stora Kopparberg combined investments in several locations with tentative efforts designed to centralize operations.

A decisive step was taken in 1868. G.A. Lundhqvist, a remarkable Mine Master and ironworks manager, spearheaded a plan designed for the radical reorganization and rejuvenation of operations. The plan encompassed all Company operations.

Efforts aimed at reorganizing the Company were drawn up by successive committees, each of which was chaired by Lundhqvist. Thus, in practice, he determined Company policy.

The situation is fully described in the minutes of the first meeting of the committee in 1868. The minutes note that Stora Kopparberg found itself in a troublesome situation and that it was now of the utmost urgency that "efficacious, forceful and swift action be taken to avoid difficulties, and to achieve a modern and more centralized management and administration."

The Company was undoubtedly in need of a more streamlined organization. At the same time, it is worth remembering that Stora Kopparberg at this stage had the largest sales among Swedish iron manufacturers and was ranked as the country's third or fourth largest company.

Increased profitability was to be gained through 'centralized management'. But it also involved an increase in output. A larger smelter works, located near Gävle and used by several companies, appeared to present an answer. This model for expansion had been implemented successfully when four smaller mills had cooperated jointly in 1856 to establish a large-scale rolling mill in southern Dalarna. Another idea, proposed as early as 1863 by one of the Company's leading men, P.A. Jacobsson, involved a plant at Domnarvet by the Dalälven river where plenty of hydro-power was available.

This must have appeared particularly attractive when the issue of an extension of the rail link between Gävle and Falun down to the west coast became acute. The idea of extending the line up to the Siljan area in the north was also beginning to be considered.

The improved communications would not only prove advantageous for sales. It would also give the proposed ironworks a substantially extended catchment area for low-price charcoal.

'A proposal to set up two blast furnaces with Bessemer converters' was recommended by Lundhqvist in 1871. A year later a decision was taken to build a complete mill, including rolling mills designed primarily for rails and heavy plate.

The plans were in line with a general trend. Besides Domnarvet, new large blast furnaces were set up at eight Swedish mills during this decade.

Stora Kopparberg intended to finance this major construction project through loans, which, however, proved insufficient. Costs were substantially higher than expected. The final bill, submitted in 1878, was SEK 4 million compared to the one million that was calculated. Also, the boom period immediately following the Franco-Prussian war had turned into a recession.

The completed mill appeared to be the result of gross misjudgement. The rescue came in the form of a series of modifications of the original design. This was the first major task of Erik Johan Ljungberg, who was appointed general manager of the Company in 1875.

The planned product program, which included quality iron for export in line with the prevailing Swedish model, had to be abandoned. Instead, the objective became one of attempting to develop a domestic customer base which required less expensive grades.

After ten years, success had at least been achieved in terms of volume. At this stage, the mill was producing 33,000 tons annually, which was approximately double the entire output of the Uddeholm mills.

However, several problems remained to be resolved, many of them resulting from the varying quality of the mill's ore base. Initially, supplies came from a number of small mines, which led to severe problems in manufacturing products of a uniform quality.

Many of these problems were solved in a radical fashion with the full-scale transition to the basic Bessemer (Thomas) process, completed in stages, beginning in the early 1890s. With his remarkable alertness to technological trends, Ljungberg had as early as 1880 established contact with the inventor of the new process, Sydney Gilchirst Thomas. Production could now be based on cheap ores with a high phosphoric content from the newly acquired sections of the Grängesberg field. The main product was to be commercial steel in a variety of forms.

All known types of steel smelting processes—Lancashire, Bessemer and Martin, acid and basic—were successively applied in the Domnarvet mill. None of these methods, however, presented as many problems as the switch to the Thomas blasting process. The problems encountered with this method were so intractable that even Ljungberg began to doubt the possibility of achieving a favorable outcome. Soon, however, Thomas became the dominant process at Domnarvet. By 1895, output had climbed to 18,000 tons. The process design at Domnarvet was remarkable in that the mill had the sole iron production facility in the world using charcoal as fuel.

The complex business of supplying the mill with sufficient quantities of charcoal was greatly simplified by the construction of a battery of eight coal furnaces around 1890. These were based on a design of the general manager himself, Ljungberg. In 1897, almost 60,000 charges of wood and laths were converted to charcoal at the ironworks at a competitive price.

Increased production at the ironworks gave rise to the need for an additional blast furnace. By this stage there was an increasing use of 'electrical power to operate machinery and, thus, the use of steam pipes and traction cables was eliminated'. Productivity in terms of tons per months had improved substantially.

Total production in 1897 was 40,000 tons of rolling iron and forged goods, which meant that Domnarvet now accounted for approximately 15 percent of the Swedish output. Apart from a major order from Swedish Railways for rails, product orientation towards primarily the Swedish market meant that the ironworks still received orders for various products in rather short series, manufactured in a large number of rolling mills. Towards the turn of the century, there were twelve of these, with a product program covering the entire range from wire and various rod products to sheet of varying thickness.

Thanks to the continuing industrialization of the country, which accelerated in the early 1890s, demand remained satisfactory. Furthermore, exports were not insignificant to such markets as Southern Europe and Latin America.

As in the case of a number of other ironworks, the production program included manufactured products. At Domnarvet, the volume of output in this respect was limited in quantity (a little more than 2,000 tons) but highly varied, which was clearly demonstrated by the exhibit at the Stockholm Exhibition. Despite the outward signs of success, the production of iron was not particularly profitable. In speeches and articles, Ljungberg often addressed the problem of poor profitability in the production of iron.

Forest industry

Discussions concerning the establishment of Domnarvet and the subsequent increasingly impressive production figures at this large facility meant that the owners and the public regarded Stora Kopparberg primarily as an iron manufacturer. The general feeling of the Company's shareholders at this time was that forest operations and the timber business were dubious and speculative enterprises.

There may have been some justification for this attitude. In many other quarters, however, there were no such misgivings. As demonstrated by the Stockholm Exhibition, the forest industries were expanding at a breathtaking pace.

In terms of volume, exports of wood and timber had surpassed the country's iron exports as early as the beginning of the 1860s. Sawn timber remained the most important export product for several decades afterwards.

The year 1897 marked a peak in export figures, which had doubled since 1872. Afterwards they remained below this record level.

On the other hand, the mid-1890s marked the beginning of a substantial expansionary period for the pulp industry.

The first groundwood mill was set up in 1857. Production subsequently expanded rapidly and was conducted at the beginning of the 1880s at some seventy mills, most of which were small. Mills with a capacity of 10,000 tons or more were not established until a decade later.

An experimental period for the manufacture of chemical (i.e. sulfate) pulp gained momentum in the 1870s.

The well-known chemist Carl Daniel Ekman was a pioneer in the manufacture of sulfite pulp. This took place at Bergvik, outside Söderhamn, at the beginning of the 1870s. The real breakthrough did not occur until the 1890s.

The greater share of the pulp was exported, but an increasing share was used in domestic paper mills, many of which directly adjoined the pulp mills.

These were based primarily on groundwood pulp. The expansion began around 1890. At this time there were 25 mechanical paper mills, against 16 purely manual paper mills. An annual production of 5,000 tons was regarded as quite high for this period.

Towards the turn of the century, a remarkable initiative was taken by the Swedish company Papyrus. In 1897 the company set up a new mill with three machines which gained "an unusually uniform and powerful development." However, modern large-scale operations were not introduced in Sweden prior to the startup of the Kvarnsveden paper mill.

A prerequisite for the expanding forest products industry was a well developed transport system which could be used to remove the raw material from the expansive forest areas to the mills. The Swedish water courses were ideal for log floating. The water courses were extended to keep pace with the expanding industry. It was estimated that in 1900 water courses on which logs could be floated had a total length of 25,000 kilometers, twice the combined length of the Swedish railroad system of the time.

Stora's forest products

Despite the doubts expressed, sawmill operations also expanded within Stora Kopparberg and began to play an increasingly important role. In actual fact, they had long formed the basis for the Company's financial solidity.

The first step towards larger-scale operations was taken by the Company in 1854, when it acquired a one-third ownership interest in the newly formed Kopparberg-Hofors Sawmill Company (later Kopparfors). Cooperation between the new company and Stora Kopparberg was

intense during subsequent years. P.A. Jacobsson, who held a leading position within Stora Kopparberg, was also general manager of Kopparbergs-Hofors from 1862 to 1874. During the same period Stora Kopparberg became increasingly involved in forest acquisitions, notably forest facilities for ironworks. The Company often adopted a rather defensive tone in describing these activities. The purpose of these operations, it was stated, was primarily to prevent speculators from ruthlessly exploiting these resources.

The logical outcome of these substantial acquisitions occurred in 1859, when a decision was made to modernize and expand the water-powered sawmill at Domnarvet which the Company had owned since the seventeenth century. In the years ahead this mill would focus on the export market.

After only a few years, production had increased to between 15,000 and 20,000 standard planks. This was usually 5,000 standard planks less than the output from the Kornsnäs sawmill, at Lake Runn, which was Sweden's largest. The raw material base to which Stora Kopparberg had access amounted to 125,000 hectares of leased forest. In addition the Company had its own forests, totalling some 50,000 hectares, which, however, it preferred to hold in reserve.

Despite the fact that the large forest companies in Dalarna had most of their forest holdings in different parts of the province—Kopparbergs-Hofors along the Svärdsjö water course, Bergslaget in Western Dalarna, Korsnäs in the Eastern parts of the province—clash of interests became increasingly common. The need for a coordinated policy became more evident.

An agreement concerning a joint organization for timber purchases was reached in 1865. Cooperation was broadened and continued at an intense pace, with favorable results up until 1889. When it was discontinued, Stora Kopparberg's holdings amounted to 123,000 hectares of forest land with ownership rights and 343,000 hectares of leased forest. By this stage the far greater share of the Company's sawing operations had four years earlier been moved to Skutskär, at the mouth of the Dalälven River.

The acquisition of attractive new leases and favorable economic conditions led to a rapid growth in production at Skutskär. During 1899, production reached 44,000 standard planks. As noted in the account of the Stockholm Exhibition, Skutskär at this time was described as the largest sawmill in the world. It was certainly the largest in Europe.

An analysis of the Company's gross profits during the last five years of the nineteenth century shows that the ironworks and mines accounted for SEK 1.2 million, compared with 1.7 million for sawmills and forests. During the immediately preceding period the difference was even more positive for forest products.

It was now time for the Company to enter the chemical pulp market. A sulfate mill with an annual capacity of 6,000 tons of pulp was constructed at Skutskär. The raw material to be used—sawmill waste—represented an innovation of some significance, particularly in terms of economy. The mill was ready for start-up in 1895. After two years of problems, operations at the mill proceeded smoothly.

The expansion of Skutskär continued with a sulfite mill. Ljungberg's interest in attaining maximum utilization of the raw material through the production of secondary products resulted in a pioneering operation for the manufacture of sulfite alcohol and other products.

In 1897 it was also decided to build what was then regarded as a mammoth paper mill at Kvarnsveden, upstream from the Domnarvet steel plant. Planned output was 30,000 tons per year, almost ten times more than that of the normal size mill of the period.

Adopting a more consistent policy than any other mill in the country, the Company focused on producing a single, but profitable product, newsprint. Customers were to be found in the expanding Continental printing industry. This was also a way to utilize spruce, a raw material that up till now had enjoyed only limited popularity and of which the Company had ample supplies.

The Copper-Mountain Works

Overshadowed by more glamorous operations related to iron and wood, the industries connected with the old mine continued to operate and at periods showed significant sales.

The eventful 1860s also entailed radical changes for the so-called Copper-Mountain Works comprising the Mine and adjecent industries in Falun. The long discussed coordination of operations now finally became a reality. Smelting was no longer a private activity reserved for the master miners. At the same time, refining was transferred from Avesta to Falun.

Mining activities were comparatively extensive. From the 1860s and into the following decade, they corresponded to one third of the record level during the Golden Age of the seventeenth century. Copper production also provided the basis for the manufacture of a range of chemical products, such as vitriol and sulfur, which provided good profitability. Red paint remained a reliable source of income. This was also a time of growing interest in the recovery of silver and gold.

Even at the end of the 1870s, sales from the Copper-Mountain Works were on about the same level as the ironworks, about SEK 1 million. In terms of its contribution to gross profit, the performance of the Copper-Mountain Works was even better, and regularly surpassed that of the ironworks up to 1880.

After this period, sales stagnated and profits declined. The role of the Copper-Mountain Works decreased drastically although operations were not insignificant. In the mid-1880s the mine and related industries provided employment for 500 workers.

In 1881, a sudden upturn occurred as a result of the new body of gold discovered by Erik Gustaf Eriksson, a small boy. The ore find consisted of quartz containing gold. After a number of years, the gold accounted for 25 percent of the Copper Mountain's sales. A certain amount of silver was also discovered. Henrik Munktell, an engineer and later general manager of Grycksbo paper mill, together with the renowned Gustaf de Laval, invented a more efficient method for recovering the gold.

Nevertheless, the continued existence of the Copper-Mountain Works was endangered towards the close of the 19th century. Important underlying factors were low copper prices resulting from major new finds in America, which depressed the market. Operations would most likely have been discontinued but for the prospect of continuing gold production.

After 1894, copper was no longer manufactured in metallic form. The age-old process, on which the Company had been founded, was now definitely a closed chapter. Metal resources were used instead for the production of vitriol. During the final years of the nineteenth century, these operations experienced a renaissance thanks to the growth of the chemical industries in Europe, and especially in Russia, which paid good prices.

The good times came to an end during the first few years of the present century. Prices fell and manufacturing costs increased. At the same time the supply of gold decreased. The Mine, however, refused to be cowed by the gloomy state of things. Sulfur pyrites, refined by a new method, offered a prospect for survival.

Power

The expansion noted in these various areas did not come to a natural end in 1897. On the contrary, the Company found itself in the mainstream of rapid development. This involved the large ironworks and the expanding industries at Skutskär.

In particular it applied to the new units that were established, especially the Kvarnsveden paper mill, where the most impressive reminder of the Stockholm Exhibition was to be saved through the purchase of the large industrial hall from the Royal Deer Park, which was to house the first battery of paper machines. The startup of the new mill proceeded remarkably fast and without hitches. A technical innovation, the production of hot-ground pulp, resulted in a highly appreciated and much demanded product.

It also involved the acquisition of Söderfors, which gave the Company access to high-quality forest and permitted a rational division of labor between Domnarvet (in its capacity as a commercial steel mill) and the Uppland ironworks, with its traditional quality profile.

The acquisition of raw material resources during the decades around 1900 was to prove of considerable importance later. Perhaps most remarkable was the acquisition of water rights, conducted on the initiative of Ljungberg since the last decade of the nineteenth century. It wasn't clear what purpose they would serve. However, Ljungberg was totally convinced that in the long run they would offer "great possibilities for the future development of Stora Kopparberg", and that it was economically advantageous to act fast. His calculations also included the benefits to be gained from control of the waters of Lake Siljan. The protracted negotiations concerning this matter began at the turn of the century. The result was that the Company found itself in an excellent position when it became neccessary to expand its hydropower facilities, a condition for survival of the Company's heavy industries during the subsequent decades.

Summary

During Ljungberg's time, the Company's sales advanced from SEK 4.5 million to slightly more than SEK 30 million. The Company had not only expanded, what was more remarkable was that its more important operations had, in most essential respects, been created during Ljungberg's period as general manager.

The structure of iron manufacturing operations had been changed completely. The ore came from mines which Ljungberg himself had acquired. The forest holdings were virtually all new acquisitions. The multifaceted forest products operation, unique in Sweden, which developed in Skutskär, as well as the large industry in Kvarnsveden, were Ljungberg's creations. Even the centuries-old processing operations at the Falun Mine had been reorganized.

Old traditions had been abandoned as early as 1868 when the Company was managed by a temporary management group led by Mine Master Lundhqvist. At long last, in 1888, a radical decision was taken to reorganize Stora Kopparberg as a limited company with a conventional general manager.

The old company was now on the way to becoming a modern, streamlined, major enterprise in terms of its organization. This also represented a formal confirmation of Ljungberg's unopposed leadership.

Socially, this was a period of great conflict in Sweden. Industrialization, resulting in large-scale population movements, created uncertainty and feelings of rootlessness among many people. New communities often consisted of little more than sub-standard barrack-like buildings.

Ljungberg and many others were highly aware of these problems, although they thought of them more as a result of past mistakes than of contemporary developments. Nevertheless, as in other areas, Ljungberg applied the same energy and acted to the best of his ability to deal with them.

The dramatic development of the country and its industries, described with such admiration in 1897, was pursued by Stora Kopparberg on an increasingly large scale in the years up to 1914.

Erik Johan Ljungberg

Nobody knows what Ljungberg looked like when he laughed. In all existing portraits he looks at us in a serious, severe and dignified manner. Nevertheless, we know that, at least during his younger years, he devoted some time to song and music and that he could display at least a superficial geniality in company.

He did not remain unaffected by the traditions of the Sweden of King Oscar II, with its idylls and romance—not least the Dalarna romance—its tassels, pomp and bunting. But most of all he was an industrialist.

Ljungberg won his position and success as a result of his intuition and sense of initiative, combined with a strong feeling for economic reality. But in particular, it was due to his tremendous industry.

When Ljungberg expressed his opinions concerning general issues, he based his position to a great extent on his own personal experience during his youth, which was marked by poverty, and the earlier years of hard work. Another frequent base of reference was the experience gained from the large enterprise he managed.

Ljungberg partly represented the old manufacturing estate manager: thrifty and cautious, well meaning in a patriarchal manner when his subordinates displayed diligence and ability, and an admirer of practical manual skills. But he was also a representative of the modern company manager: technically aware, prepared to accept radical change and invest wholeheartedly in new products, or in old products on a different scale, unfettered by tradition.

For him, the value of work was always a central conception. For him, assiduous effort was the main key to success.

Ljungberg's long-time friend Richard Åkerman wrote in an obituary: "Nobody was more certain than Ljungberg that a country's well being depends on the efficiency of its inhabitants and on the average units of value achieved by each individual in relation to the time spent on his work." Ljungberg formulated his ideas in a similar manner.

In a speech concerning productive labor, he said that he had estimated that the total production of the Swedish nation amounted to SEK 409,675 per hour, and continued: "This is the amount that is wasted for each hour unused by the country in productive labor". And he continued with his well known and often repeated maxim "Time is worth more than money, since lost time can never be regained, while lost money can." There was only one way to improve the economic conditions of the people: "The utilization of our country's economic resources through the creation, encouragement and support of all productive labor". On the other hand he cited as counter-productive legislation resulting from "misguided, social emotion". The objective was a population which was "well prepared to fulfill its duties and prepared not just to live its own life in the happiest manner, as permitted by conditions, but to contribute to the increased happiness of future generations."

More work would provide an increased degree of happiness. "If our workers were aware of the situation of industry, I believe that they would not ask for increased benefits without a corresponding increase of the products of labor and if they themselves knew what really fast, skilled workers could willingly and happily accomplish, their greatest desire would be to acquire a corresponding capacity, that their services in our home country would be equal to that of the Americans."

There was, Ljungberg believed, extensive scope for greater effort. As regards enthusiasm we trailed far behind other countries: "Our country is not poor, but we are generally too lazy."

It is said that Ljungberg was sceptical about theoretical knowledge and had little time for people with academic qualifications. The claim is not without a basis in thruth, and one of the underlying reasons was perhaps that the technical colleges of the time did not have the prestige they later acquired and that industrial operations in the factories during this era had a greater use for a skilled foreman than a civil engineer.

Furthermore, Ljungberg claimed, technicians lacked of any sense of economics. "It is extremely difficult," he says "to find business-oriented people with an engineering qualification.They usually think and act on the basis of their technical thinking and forget about the accounts. Senior management must be in the hands of a business-like, calculating administrator".

His attitude was most definitely not hostile to technology. In an introduction to a survey of industry and technology he states that "technical progress in recent years is so great, so pervasive in many areas, that the miracles described in our beautiful sagas resemble small trifles". And he continues: "Technology has been an invaluable handmaid to industry and it is only through its great contributions that the varying industries of the world, its communications above and below the soil, above and below water, and in the air, have been achieved. Through these contributions, wars have been curtailed, defense has been made more effective... Thanks to the appropriate utilization of these contributions it has been possible for industry to grant, within certain limits, the higher wages demanded by manual laborers."

The task of science was a similar one. His opinions on this subject are expressed in a speech in connection with his election to the degree of Honorary Doctor. According to Ljungberg, in times past, knowledge of the laws of nature had been the preserve of a few people who thereby ensured the admiration of others. Now science had a loftier objective "since now, hand in hand with research, there is a lively outward-oriented effort, an effort to spread the fruits of science to achieve the maximum benefit for mankind".

"The lantern of science is needed in all areas of life, from the cradle to the grave, and it is a joy to witness the work in progress, aimed at spreading a healthy light". As an example, Ljungberg mentions the application of science to industry, which utlizes knowledge of chemistry, physics and electricity. The same applies to agriculture, to the home (where the kitchen had been transformed into a 'chemical-physiological laboratory'), and to schools and medical care.

Ljungberg's interest in the upbringing and education of children resulted in his appointment to a government committee for the organization of public education.

His opinions in the matter were very clear. He believed that education should be designed to impart practical knowledge which could be utilized directly in the student's life, regardless of whether the student became a skilled and diligent housewife or a skilled and diligent factory worker.

America was the model. There, people were "trained for intensive labor in response to the prevailing social conditions. The unmanly emotional whining in our country is absent over there. A manual laborer wants to be a man, a real man and not a poor wretch who has to be taken care of and pitied".

His objective was to plead for the type of practical education he himself received and which contributed to his success. He had nothing against the advancement of energetic and skilled people.

However, in all his benevolence he had little understanding of the workers' desire to determine their own destiny.

Ljungberg made major contributions to improving the living conditions of his employees. He had houses constructed which, given the conditions of the time, were of a high standard.

He set up bath houses, medical facilities and sanatoriums, although one must admit that he never lost sight of the ultimate objective, namely, to improve labor efficiency.

In a similar fashion he wanted to justify his considerable contributions in the area of education. In 1892 he organized peripatetic cooking courses for housewives. Improved nutrition would produce improved working results. A few years later he decided to set up a technical school at Domnarvet. From this beginning, there developed the highly varied activities of the company's Vocational Schools, with educational facilities in a range of technical subjects based at Domnarvet, Kvarnsveden and Skutskär.

For future housewives he created a corresponding type of school in five different locations within the area covered by the Company.

Ljungberg and his wife left their entire fortune to the schools. The purpose is formulated in their will in which they expressed the hope that the schools "to an effective degree" would "contribute to creating a strong, intelligent breed of workers, healthy in body and soul, and equally skilled, strong, healthy and intelligent women capable of fulfilling their duties and thus building happy homes with happy inhabitants".

Ljungberg was not an original thinker. On the contrary, what is interesting about his ideas and speeches is that they are so representative of the mentality of a major industrialist at the turn of the century.

Many of his ideas were common knowledge, probably originally gleaned from a didactic philosopher such as Condorcet or an early nineteenth-century thinker such as Saint-Simon.

Sometimes he displays a somewhat bewildering combination of old-fashioned patriarch and new rational man. Even in 1923 Gustaf Dalén, a Swedish industrialist, confirmed that the vast majority of his contemporary entrepreneurs were of a patriarchal bent.

The passion for work which was preached by some industralists in Sweden knew no bounds in America. There, productive work was regarded not only as the source of all well being, it quite simply represented the true purpose of life. An American phenomenon which Ljungberg could hardly avoid being aware of was the Knights of Labor, who experienced their heyday during the 1880s, at which time the membership of the organization easily surpassed that of the conventional trade unions. The Knights of Labor praised sobriety and the ideals of the Free Church, but focused on work, which they regarded as the very foundation of society, art and science, the base that permitted the realization of God's intentions.

These ideas meant that the worker and the industrialist were really on the same side at the barricades. On the other side stood the speculators, the contemptible 'jobbers', to use Ljungberg's expression.

In America, where there were no barriers to industrial and technical development, progress proceeded rapidly. "The old nations of the Earth crawl forward at a snail's pace, but the Republic speeds past them like an express," wrote Andrew Carnegie, one of Ljungberg's mentors.

Belief in progress and a better world were the emotions which characterized public thinking. Technology continually created new wonders. The industrialists were the heroes of the day. The machine was their symbol, thrift and diligence were the cardinal virtues. "In the industrial society", wrote Herbert Spencer, who was originally a railroad engineer, "prejudice, superstition, violence and ignorance will be exterminated". In his much read work on the development of civilization in England, Buckle, 'the Darwin of history', represented a simple and unshaken belief in mankind's march towards increasingly higher and clearer horizons.

Samuel Smiles, a highly popular author of his time, expressed ideas of which many are repeated by Ljungberg. In his writings, Smiles tirelessly praised work, science, technology, the will to progress, and public education. As in the case of Ljungberg, it was an advantage if one started one's career from the bottom. Men such as James Watt and George Stephenson were excellent examples. These were men whose success was based on their own strength. Work and thrift were the only ways to happiness. This is what should be taught in schools, instead of devoting one's energy to play. Smiles most successful book, Self-Help, was first published in 1859. It was translated into at least 17 languages. Its Swedish title read 'The inner strength of Man—the right way to fame and riches', and was published in eight editions, the final one appearing at the beginning of the present century.

The theme pursued by Ljungberg in the speech he gave at his doctoral ceremony, concerning the practical benefits of science, could hardly be called original. A widely disseminated work on the subject, written by John Herschel and entitled 'On the Study of the Natural Sciences' was included in the well-known Swedish 'Library of Popular Natural Science' published in several volumes during the 1840s. Similar ideas were also expressed by the great prophet of positivism, Auguste Comte.

The attitude to life represented by Ljungberg and many contemporary opinionators no doubt appears naive today, presenting an over-simplified and schematic view of people. The idea that development in one way or another has its price was completely foreign to them.

Their successors have learned that the interacting forces of existence create a much more complex and contradictory situation than they ever imagined possible. A better world requires more than good will and diligence—still they are characteristics that can still get us a long way.

A new direction

EACH YEAR, the General Secretary of the highly respected International Iron and Steel Institute publishes a report on the situation within the international steel industry. In the report distributed in autumn 1974, he stated, on the basis of extensive data, that the most crucial problem for the world's steel manufacturers was how to cope with the continual increase in international demand, a demand which was evidently going to remain acute. Expansion would be required worldwide. The capital requirement was enormous. Capital expenditures to solve the energy problem would also be an issue.

Seldom has a forecast been more quickly and emphatically refuted by reality than this one. In the case of steel, the turning point occurred towards the end of 1974.

During the spring of 1975 all the industrial countries in the West were heading into recession, leading to inflation and unemployment. The fall in steel prices during the latter half of 1974 and up to March 1975 was the most substantial since the beginning of the Depression in the 1930s. Not one single steel manufacturer in the EEC was able to report a profit or even break-even. Following this period, the downward slide became increasingly steep.

The year 1977 marked a nationwide low for the Swedish economy. This was the year in which the Swedish krona was devalued on two occasions. Gross national product decreased for the first time since the Second World War. Swedish exports decreased, imports increased. The newspapers were filled with reports of new crisis-ridden companies. The situation was described as 'a Swedish industrial tragedy.'

The implications of the unfavorable conditions were not limited to the economic arena. They also gave rise to a crisis of a psychological nature. Employees who had felt that they were making a socially useful contribution in large, financially strong companies, with recognized quality products, suddenly found that the basis of their existence had been removed. The products that they helped to produce no longer had a buyer, there was no use for their labor input. Since everybody was so ill prepared the experience came as an even greater shock .

Worried industrialists literally queued up outside the door of the Swedish Minister of Industry. Radical measures were required to rescue substantial sections of industry. Many companies underwent fundamental changes.

This applied in no small way to Stora. Starting in 1974, a period of radical change was initiated which culminated in 1977. At the start of that year, an agreement came into effect whereby the Company transferred its specialty steel operations to Uddeholms AB. A few months later, in the same year, a government survey of the situation within commercial steel operations was published, containing an analysis of the industry and proposals for its future. Based on the results of the survey, intensive and complex negotiations were conducted among the owners of the three largest commercial steel mills—Stora with Domnarvet, the Gränges company with its large and newly built steelworks at Oxelösund on the Baltic, and Statsföretag (a Swedish state-owned holding company) with its steel company, Norrbottens Järnverk (NJA), located at Luleå in the far north. Several other parties, not least the Government and the trade-unions concerned, participated in the negotiations. The final result was the formation of SSAB, Svenskt Stål Aktiebolag (the Swedish Steel

Company), which formally started operations at the end of 1977.

As a result of this decision, which meant transferring control of the remaining steel operations of Stora, as well as its iron ore mines, the Company's role as a steel manufacturer was now a closed chapter. The Company's operations had virtually been halved.

Parallel with the phase-out of the Company's steel operations, the forest products sector received substantial reinforcement. During 1977, the first stage of a major expansion program for the pulp mills at Skutskär was finalized. It cost SEK 700 million.

This marked a radical reorganization of Stora's entire production of pulp. Sulfite pulp production at the Company's mills in Kvarnsveden, Grycksbo and Skutskär was discontinued. Expansion focused on bleached sulfate pulp. The manufacture of chemical pulp within Stora had focused on a single streamlined, energy-efficient and environmentally satisfactory unit, one of the largest in the country.

Paper operations expanded when the Company raised its joint interest in Hylte paper mill, in southern Sweden, from 10 to 20 percent. Stora had provided part of the capital stock and acquired responsibility for some of the company's product marketing in 1970, when it was reorganized, and a decision was taken to build a newsprint machine at the mill.

But the major step in forestry operations had occurred in 1976, when the power and forestry company, Bergvik and Ala, was merged with Stora. This was a considerable success for company management. "It is certainly no exaggeration to say that the acquisition of Bergvik and Ala is the most significant improvement in Stora Kopparberg in many decades," said Erik Sundblad, the Company's managing director. Stora was now well on its way to merging the operations of the two companies. Sundblad's ideas concerning the future of the newly acquired units was crystallized in the not easily realizable plan of using the existing industries as a base, to create a large, wood-consuming industry at the mouth of the Ljusnan River, which would operate in symmetry with the complex at Skutskär, at the mouth of the Dalälven River.

Combined, these measures meant that at the end of 1977 Stora's corporate profile was largely that of a forestry and power company.

The changes followed a pattern which occurred in many places. The traditional type of company, an industrial concern comprising forest activities, iron manufacturing, power production and, sometimes, agriculture, controlled by a single management, was rapidly dying out throughout the country.

The Old Company

For many people, what was happening to Stora was frightening and difficult to comprehend.

For decades, people had regarded Stora as a permanent feature of the Swedish industrial environment. It was taken for granted that it was and would continue to be Stora's natural task to spearhead the refining and processing of the natural resources of Dalarna and the Dalälven River basin. The ironworks and the paper mills of Domnarvet and Kvarnsveden, the mines in southern Dalarna, the sawmill and pulp mills at the mouth of the Dalälven and the specialty steel mill just upstream at Söderfors, the power plants on the Dalälven, and agricultural holdings, which, combined, covered more space than any other in the country; all this was considered as constant.

Thanks to its multifaceted character, its ability to act independently, and the fact that it was older than the kingdom of Sweden (and much more solid), it was not unusual for the Company to be regarded as a type of principality by employees and outsiders alike.

There is no doubt that the Company's management and many of its employees felt that they were working for a Company which was in a class of its own. "It's amazing how conceited all those who have to do with Stora become," wrote Marcus Wallenberg (Senior) on one occasion.

The Company's image of power and wealth was undoubtedly helped by the taciturn and withdrawn nature of the Company's managers since the early 1920s. A typical example of this is a memo from 1944 in which the general manager Ejnar Rodling declares "it is not in the interests of Stora Kopparberg that matters affecting the Company become public knowledge". Thus, comments in this respect "should be avoided to the greatest extent possible". More or less the same principles governed the form of the Reports of the Board of Directors of the Company.

Instead, the guiding star for all Company general managers was to act covertly. Thrift, cautious expansion and safe transactions were the key-words.

The first representative of this corporate philosophy was Emil Lundqvist, who was appointed in 1923 and managed the company for almost twenty years. He came to the Company preceded by his reputation as an energetic resolute corporate trouble-shooter. Working intimately with the Chairman of the Board, Marcus Wallenberg, he guided the Company with a steady hand through fluctuating economic cycles.

The dominant problem in the 1920s was the Domnarvet ironworks. The favorable conditions during the war and the short-lived boom immediately afterwards had encouraged management to draw up large-scale plans. The need to face reality was painful. For the first time in its entire history, the Company was forced in 1922 to suspend its dividend payment.

For Domnarvet, the trend became increasingly ominous. Prices for commercial steel continued to fall. In the words of Lundqvist the market situation was "thoroughly rotten". Domnarvet reported million-kronor losses several years in succession. Generally speaking, however, the performance of the ironworks was not much worse than that of other Swedish plants in the same industry. Production was maintained at a relatively high level. The production of billets during the first half of the 1920s accounted for an average of one seventh of total Swedish output. During the last five years of the decade the situation became more favorable. The construction and engineering industries within the country began to gain momentum. The number of customers increased. Furthermore, in the mid-1920s, Domnarvet received a new, determined general manager in the form of Claes Wahlund.

Remarkably, the deep depression around 1930, which halved Swedish exports and resulted in a decline in Swedish GNP of about 20 percent, had only a slight effect on Domnarvet.

The favorable trend continued. The situation was assisted by the international situation, which was fraught with tension and which, in turn, resulted in moves towards an arms buildup. In 1936, Domnarvet reported that, despite substantial expansion of existing facilities, it could no longer completely meet demand but would have to limit itself to "meeting the most acute requirements of its old customers".

Towards the end of the decade, Domnarvet posted a profit which was twice as large as the combined profit from the Company's two large forest-industry units, Skutskär and Kvarnsveden. Any talk of a possible closure was silenced.

However, the fact remained that both Wallenberg and Lundqvist were in principle sceptical about commercial steel operations in Sweden. An expression of the lack of faith in this industry came in the mid-1920s when Lundqvist sold a large holding in the Grängesberg field, Domnarvet's most important raw material source. The measure has been critized. Lundqvist's argument was that the remainder of the holding should suffice for the foreseeable future. Moreover, at this time the Company needed to strengthen its liquidity.

The industry which played a primary role in contributing to the economic boom in Sweden during the mid-1930s was, however, the forest-products industry, notably pulp and paper.

For Stora, the pulp mills at Skutskär were a central concern. They represented by far the largest investment object within the Company during this decade. The result was a new, and soon classic, bleached sulfate grade, designated STORA 32. Intensive development work—located at a central laboratory in Falun since the 1930s—was also conducted on sulfite grades but did not lead to any immediately practical results.

New secondary products saw the light of day. An epoch-making process for chemical recovery was developed. Much of the honor for this is due to Sixten Sandberg, technical manager of the Skutskär mills and later the founder of the central laboratory. As an innovative engineer he was to play the role of mentor to an entire generation of Swedish pulp engineers.

The Company's most reliable profit center was, since its establishment, the Kvarnsveden Newsprint Mill. A modernization program, which produced no major capacity increase, was completed during the 1920s. At the beginning of the next decade, despite indications of an impending recession, Lundqvist decided to make a major investment in the paper mill, the first in 25 years. A new machine, the largest in the Nordic region, was started up in 1931 and brought Kvarnsveden's capacity up to 100,000 tons per year.

Total production volume within the Company's forestry industries did not, however, increase very much. Consequently, there was no need to increase the amount of forest holdings. Lundqvist declined to purchase forest land even when the opportunity presented itself.

On the other hand, the power division experienced a major expansion. The Company, which was already a substantial producer, with four large power stations within the city limits of Borlänge, now opted for what was then a 'gigantic' power station downstream in the Dalälven at Långhag. It was from here that the water level in Lake Runn would in future be controlled. Långhag was completed in 1939.

Stora had been the first Swedish company to use hydropower on a large scale for its industries. One adverse affect of this was that in the alternating-current generators built in the 1890s, based on an American model, the Company opted for a periodicity of 60, which was almost unique among producers in Sweden. To achieve coordinated operations with other producers it was necessary to conduct an extensive rebuilding program over a period of 20 years.

The Falun Mine was developing favorably. The mining of sulfur pyrites, which formed the basis of operations, was almost 100,000 tons per year in 1940, almost double the output from 20 years earlier. The demand for sulfuric acid was increasing, and by the end of the period the increase was almost five-fold. In addition, there was also the manufacture of red paint and vitriol.

In 1926, a notable development was introduced, when the old wooden constructions were gradually replaced by stainless steel pumps and piping, capable of withstanding the corrosive water. Apart from this, investment at the Falun industries focused primarily on the sulfuric acid plants.

It was only natural that someone of Lundqvist's mentality should be sceptical about the broad social involvement which formed a traditional part of the operations of a company like Stora Kopparberg at this time. Stora's involvement in this respect was probably greater than that of many comparable companies, due not least to the measures introduced by Ljungberg.

The supply of housing was one. At an ironworks like Söderfors, almost all employees lived in buildings owned by the company. Lundqvist was forced to expand such activities to a certain degree. This applied in particular to Domnarvet, where the housing shortage, especially towards the end of the 1930s, was acute.

At the beginning of the 1930s the Company continued to operate some twelve food stores at small and larger ironworks. These were phased out about 1932, and the very extensive medical care apparatus which had been built up at Domnarvet was reduced at the same time. Lundqvist maintained that the Company's task was to manufacture iron, not heal the sick. The vocational schools set up by Ljungberg, were allowed to continue their activities at approximately the same level.

Looked at in perspective, the 20 years of Lundqvist's stewardship might appear to be rather uneventful. His contribution is probably best illustrated by the tables showing the Company's sales, profits and depreciation. All of them indicate increasing figures. Net sales increased threefold, to SEK 131 million; profit increased from slightly less than SEK 2 million to more than SEK 11 million. At the same time, annual depreciation exceeded net profit. The value of fixed assets was written down from SEK 58 million to SEK 17 million. Consequently, Stora became known as a company of unique solidity.

Despite the favorable development, shareholders did not receive much. The dividend remained at 6 percent over a long period and did not increase to seven percent until the mid-1930s. A larger increase, to 12 percent, was made during the war boom in the 1940s.

A time of prosperity

During the period from the end of the Second World War and up to the mid-1960s there was a sharp increase in industrial production throughout Sweden, not least within the paper and pulp industries.

Capital expenditure was substantial, notably during the latter half of the 1950s. The initial years of the following decade resulted in record profits. This applied in particular to shipbuilding, as well as to the steel industry and mining. The highway network was expanded and improved, housing construction boomed, and cars became increasingly common. These provided the conditions for an expansion of the engineering industry.

Much of this was due to the internationalization of trade, the establishment of the EEC and EFTA and, in Sweden, to labor market policy which facilitated the flexibility of labor.

Within Stora Kopparberg, Lundqvist's successors, Ejnar Rodling (1943 - 1948) and Håkan Abenius (1948 - 1966), had the same ambitions as the original corporate troubleshooter.

They were well supported by the leading members of the Board: the chairman, Robert Ljunglöf and, from 1950, Jacob Wallenberg and later his brother, Marcus Wallenberg.

When the Second World War began to draw to a close, there were plans to expand virtually all of the Company's operations. These included major investments designed to double capacity at Domnarvet, uprate the specialty-steel production program at Söderfors and, to support these measures, to intensify operations at the Company's ore mines. Stora wanted to modernize the Skutskär sawmill and further improve the quality of the pulp products. Expansion at the industries required more energy. The Company still held old, unutilized water rights, and new ones were to be acquired. Operations affected the entire Dalälven river system.

Large sections of the Company remained profitable without any immediate major investments; Kvarnsveden continued to be a reliable profit center. The chemical industries in Falun also yielded a good return. The forests were regarded as an asset that grew of its own accord.

It has been claimed that investment in Sweden could have been greater during the boom conditions of the 1950s. Stora can hardly be accused of neglect in this respect. The Company's annual investments increased three-fold during the period 1945 to 1950 and by more than 100 percent during the following decade up to 1958.

When Abenius glanced back at his ten years as general manager, he could confirm, with some pride, that while industrial production in the country as a whole had increased by 32 percent during the period, the increase within Stora Kopparberg, computed in the same manner, was a full 78 percent.

The background to the noticeable increase primarily derived from the extension of Domnarvet, which enabled the ironworks to attain a capacity of 500,000 tons per year by 1959. Thus, Domnarvet could without exaggeration be described as a completely new mill compared with what it had been at the end of the war.

As in the past, the objective was to supply the Swedish market with virtually all the commercial steel products it required, from wire to heavy plate. Furthermore, wide ranging research work had been completed, one of the results of which was the so-called Kaldo process, the Company's own, new method for the decarburization of crude iron, using pure oxygen in a rolling furnace.

The rising demand for ore, primarily the result of the extension of the Domnarvet works, resulted in an increase in mining operations in the Company's mines. Hydropower production also rose. At the end of the 1950s production had reached 2 billion kWh per year. At the same time, it was possible to forecast dramatic growth in the years immediately ahead, when the power station on the upper eastern arm of the Dalälven River at Trängslet, with its colossal stone and earth-fill dam, was due to start operation. According to plans, this was to be followed by four additional wholly- or partly-owned stations.

At Skutskär in the mid-1950s, the Company launched a sulfite pulp based on pine and produced by cooking on a sodium base. This represented the practical application of a protracted and extensive research project begun by Sixten Sandberg. In 1958 it was decided to double the capacity at Skutskär to 200,000 tons per year. The production of both sulfate and sulfite was to increase.

Without having to make any major investments, it proved possible to raise production substantially at Kvarnsveden by means of rationalization measures. At the beginning of the 1950s a new project was planned, resulting in 1956 in the start-up of an additional paper machine. At this stage capacity tonnagewise was almost on a par with that of Skutskär.

The specialty steel mill at Söderfors gave no great cause for joy, but production increased, particularly after the start-up of a new sheet mill for stainless strip in 1957. The production of bar steel stagnated, as did that of cemented carbide, production of which had started in 1944.

The development meant that the Company's total sales at the close of the 1950s were about SEK 600 million, while ten years earlier the corresponding figure was slightly above SEK 200 million. At the same time, operating income had increased four-fold.

For Stora, the initial years of the 1960s meant that the level of annual investment, almost without exception, exceeded SEK 100 million per year. Sales were approaching SEK 1 billion. But with the exception of a single year, operating income was hardly more than that attained in 1958.

The most spectacular increase within the Company during the eight years up to 1966 was achieved through the expansion of power facilities. When the Trängslet power plant was completed in 1960, the company at once increased its production by about 700 million kWh per year.

Decisions and measures of a decisive importance for Domnarvet were adopted at the end of the 1950s, in the form of major investment aimed at the manufacture of thin steel sheet. This was intended to be the main product of the mill in the years ahead.

The core of the plant consisted of a rolling mill for broad strip, backed up by a cold rolling mill. Both were completed in 1962. At the same time, the quality of the finished product was increased by galvanizing an increasing portion of the thin steel sheet.

The extension also meant that Domnarvet could roll stainless steel strip from Söderfors, which was then cold-rolled in the smaller mill. The result was that Stora Kopparberg could deliver stainless strip in broader dimensions than any other company in Europe.

During the first half of the 1960s, Skutskär entered a development phase not unlike the feverish activity in Domnarvet.. Expansion projects succeeded one another at a fast pace.

Among the comparatively few initiatives outside the framework of traditional operations was the establishment of Koppartrans Olje AB (an oil import and refining operation) in 1947, in cooperation with Rederi AB Transatlantic, a shipping line. The background to this development was the enormous growth in the consumption of oil in Sweden after the end of the war. The Chairman of the Board of the shipping line was Robert Ljunglöf, who, as already mentioned, held the same position within Stora Kopparberg. Koppartrans developed rapidly and by the beginning of the 1950s was the country's largest refinery. But as far as Stora was concerned it was merely an episode. The Company sold its share in 1963.

During the 1960s Stora acquired three smaller mill operations: Stjernfors-Ställdalen in 1961; Vikmanshyttan, and Grycksbo in 1966. The primary purpose was to be able to increase the forest holdings. Ställdalen included a varied but small-scale forest industry. The dominant industry at Vikmanshyttan was the quality steel mill, while at Grycksbo the paper mill was the main enterprise.

The sale of the shareholding in Koppartrans was related to what in the long-run was the most dramatic and in many respects the most fateful initiative taken during Abenius' period as general manager, namely, the decision to build a pulp mill in the Canadian province of Nova Scotia.

The reason for establishing an operation outside Sweden was that—according to the Company's view—there would soon be no possibility of expanding the Company's operations within the domestic forest industry. If the Company was to retain its market share, then a foreign venture was unavoidable. Along the southern part of Canada's Atlantic coastline were substantial forest reserves located at a favorable distance from both American and European buyers. The provincial government was positive to the proposal. The supply of labor, even if untrained, was more than adequate. A deep-water harbor, ice-free all year round, was also available.

Nevertheless, there were reasons for the nagging doubts still experienced by the Company's management. The venture involved the development, primarily through borrowed funds, of an industry highly sensitive to changes in the business cycle. Furthermore, operations were to be conducted in an area which almost totally lacked any industrial tradition.

The Company 's Board of Directors made their decision in 1959. When operations started up two years later, a number of advantages became evident. The Company had gained an American partner who put up 20 percent of the capital stock and agreed to purchase a substantial share of production. The supply of raw material started to flow quickly and efficiently. The mill was able to start operations without any major problems and functioned satisfactorily. However, what failed to function was the market. Prices had fallen dramatically during the construction period, and the mill reported consistent annual losses. Year after year, the parent company was forced to send funds to Nova Scotia.

Two weaknesses became increasingly obvious. One was the decision to manufacture sulfite pulp, which was considerably more difficult to market than sulfate. The other was related to the size of the plant. When planned, the plant appeared to be sufficiently large, but when it came on line it proved too small in relation to the high overheads that had to be covered.

In the last annual report to be signed by Håkan Abenius, the Company noted record sales of SEK 940 million, and a reasonable profit of SEK 42.5 million.

Expansion continued on almost all fronts. From the inside, however, the situation seemed fraught with problems. The reasons were increased costs, declining prices and production stoppages, not least at Domnarvet. The Company's shares had fallen continually during the year and were about SEK 100 lower at the end of the year than they had been at the beginning.

In an interview prior to his retirement, Abenius referred to his last year in the Company as the most difficult during his career as general manager.

A new situation

The problems were not Stora's alone. They were nationwide.

During the latter half of the 1960s, output in Sweden continued to increase at about the same pace as previously, together with investment, but the situation for industry was not unequivocally favorable.

The were several reasons.

The countries ravaged by war had by now fully rebuilt their manufacturing apparatus. Trade had become internationalized to a considerable extent as a result of cheaper long-distance transport. Proximity to the market no longer offered the same advantages as before. This created problems, especially for Swedish ore exports. Japan started to look like an extremely serious competitor in industries such as shipbuilding, steel production and engineering.

The newly industrialized countries encountered increasing success in marketing their labor-intensive standardized products.

There were also increasing problems within the country. The wage level was being pushed upward and surpassed that of many competing countries. Attempts to help low-paid workers, security of employment legislation and the greater number of employed women meant increased resistance on the part of labor to job relocation. Profitable industries sometimes encountered difficulty in recruiting personnel. Public sector consumption had increased and was now higher in Sweden than in any other country. The tax burden increased. Demands for investment in environmental programs were lobbied more powerfully in Sweden than in many other countries, imposing costs on Swedish industry which foreign competitors were to a great extent able to avoid, at least for the time being. Interest on loans increased and created serious problems for companies which had earlier acquired large debts during more optimistic times.

The result of these increasing difficulties meant that companies redoubled their efforts to reduce costs rather than expand. Labor-saving equipment was purchased, the product range was trimmed. Less profitable companies were closed down or were merged with more efficient ones. New company start-ups were less frequent.

It was during this time of change that Stora acquired a new boss. In the fall of 1966, Erik Sundblad was appointed president of the Company.

From the moment of his appointment, Sundblad clearly regarded the multifaceted operations of the Company as a problem. He was of the opinion that it was necessary to attempt to concentrate manufacturing on products in which the Company had special expertise and was therefore most competitive. The Company should also attempt to abandon standard qualities for bulk goods which were sensitive to changes in business conditions. But this was a long-term issue.

Another problem was that the organization in his view, was ineffective. The existing system meant that each manager of a mill or a division of any importance was directly subordinate to the general manager. In the 1960s, this meant that about twenty managers presented their problems and received their instructions directly from the managing director.

At least in theory, Sundblad radically reduced the number of administrative routines by combining forests and forest-product industries to form a single sector with a joint manager, with mines and steel mills forming another sector.

Steel mills and mines

The depressed economic conditions did not immediately have a specially serious effect on Stora Kopparberg. The Company still had the financial freedom to strengthen its position.

Expansion programs already embarked on were completed. In certain cases they grew to become more comprehensive than originally planned. Furthermore, a number of new initiatives were taken.

Domnarvet found itself in the middle of an intensive period of investment. According to Sundblad, this was "a good mill, fully competitive with much larger foreign steel mills".

According to a long-term plan—the first version of which was finalized in 1968—designed to present a unified picture of desired and probable development within the Company's entire steel sector up until the end of the century, the forecast for Domnarvet predicted an increase in the capacity of metallurgical and rolling-mill operations, with the focus on thin metal sheet.

The plans for the metallurgical area included a transition from the Kaldo process to a new, more flexible steel-smelting process, which resulted in increased interest in low-phosphor ore.

Production continued to increase. In 1972, the plant's rolling mills totaled about one million tons of finished materials.

The growth of Domnarvet assumed the availability of increased quantities of ore. This caused problems. Apart from the field in Grängesberg, the Company's ore bodies were comparatively small and provided only limited scope for improvement. Sundblad took a radical step to increase the ore base in 1974 when he completed the purchase which made Stora the sole owner of the large Dannemora mines. A decisive reason for the purchase was the extremely low phosphor content of the ores, which was ideally suited to Domnarvet's new metallurgical needs.

A significant factor concerning the question of how the Company's new steel processing should be further developed was the close cooperation between Stora and The Grängesberg Company since the 1960s. From the very beginning, this cooperation applied at all levels of operation, from ore supplies, product programs and investment plans to sales and their organization.

Even if discussions did not lead to a total merger as was periodically expected, the results were nevertheless substantial. This contact, which according to Sundblad was the most intimate relationship he had come across between two independent companies, continued in a spirit of harmony. For the Company it was "very productive and very pleasing". One knew, Sundblad said, "everything about each other". The close cooperation proved of value, not least when the entire Swedish commercial steel industry found itself facing a major crisis a few years later.

At the specialty steel mill in Söderfors and at the newly acquired Vikmanshyttan works, development presented more complex problems. The manufacturing program was divided up between the two mills. This was not carried out without a certain amount of friction.

The most important product was stainless sheet, manufactured in cooperation with Domnarvet. Other products included speed steel and high-alloy tool steel.

The manufacture of cemented carbide posed problems of its own. When manufacture started after the war it appeared to offer excellent prospects. The Company had access to first-class technical expertise, but it never proved possible to achieve volume production. Now the Company was faced with the choice of investing heavily or divesting itself of the entire operation. When Sundblad received an attractive offer from Fagersta, he had no reason to hesitate. The transfer took place in the fall of 1968.

Despite the efforts to rationalize operation the income from specialty steel continued to be unsatisfactory. One loss year was followed by another. A little consolation was gained from the fact that the mill shared the same fate as other companies in the industry. The main reason, management believed, was that a number of new foreign competitors had grown up who were capable of operating on a larger scale and often supported by a substantial domestic market.

The good news for Söderfors came in the form of a new grade of extremely high-alloyed tool steel, manufactured according to a powder technology process in cooperation with ASEA. The new product, designated ASP-Steel, was a success, at least in terms of prestige.

Forest industry

Sweden's forest industry constituted one of the business sectors which performed well even after the mid-1960s. This sector of Stora operations maintained a high level of activity.

An expansion of the pulp mills at Skutskär was almost complete by Christmas 1966. The most important new unit was a continuous-process sulfate digester.

A new, large-scale paper machine was installed at Kvarnsveden. In reality, the project involved no less than the establishment of a totally new mill alongside the old one. The largely Finnish-constructed plant was opened by President Kekkonen of Finland in 1969.

The investment proved successful. Operations went smoothly. A program to modernize Kvarnsveden's older machinery then followed. The aim of this project was to produce higher quality grades than ordinary newsprint.

The newly acquired Grycksbo mill was also modernized. The objective in this case was to concentrate on special high-grade printing papers. The sales organization was extended through cooperation with others, notably Papyrus.

A significant step in strengthening the Company as a manufacturer of printing papers was the acquisition in 1975 of the printing ink manufacturer G-man, in Trelleborg, at the southern tip of Sweden. The interplay between paper and ink could now be studied in greater detail than had previously been the case.

The situation for the Company's forest industry in Nova Scotia, however, was almost catastrophic. Each year brought new losses. If the operation was to continue at all—and at this stage the Board were in some doubt on this issue—it would be necessary to expand the mill substantially in one way or another to generate enough income to cover the costs of the initial investment. This was Sundblad's first major task. After two years work he made some concrete proposals in the form of a long-term plan. This involved expanding the pulp mill and adding a newsprint mill. The machinery was to be similar to that used in Kvarnsveden.

The plan was accepted. The expansion was completed almost on schedule and the new units could be started up in 1971 and 1972. After another year the Nova Scotia mill was able to report a net surplus, although a very modest one. The remarkable stamina displayed by the Company's management was at last beginning to pay dividends. Towards the close of the 1970s Sundblad could describe the Canadian mill as a "healthy and sustaining member of the family". By this time the mill was on the way to becoming one of the Company's most profitable units.

Forestry

The fact that Stora was capable of maintaining its ranking as one of Sweden's top companies during the recession was due largely to its strong base in terms of ample resources: forests and power.

Forest resources had increased substantially as a result of a number of acquisitions, some of which have been noted above. Operating methods underwent a radical transformation. In 1946 the company bought its first road-building tractor. From 1950, mechanization gained momentum and increased rapidly both in terms of felling and transportation. Soon, each year witnessed substantial investment in new machinery. In this area, Stora was one of the leading companies in Sweden.

Between 1967 and 1970 it proved possible to reduce the workforce by 50 percent. Personnel reductions continued after this period but at a less dramatic pace.

The Dalälven River's role as a means of transport changed. Throughout the 1940s and for a number of years afterwards it had been the country's most important logging river, bearing almost 30 million logs a year. As late as the mid-1960s, it carried an annual volume of more than 20 million logs to the mills, but subsequently declined rapidly in importance. By 1969 all wood supplies to Kvarnsveden arrived by road. A year later, log-floating had ceased completely. The traditional method was too labor intensive and also tied up a substantial amount of capital. It was replaced by a combination of road and rail transport.

Towards the mid-1970s, the question of the Swedish forest industry's "timber slump" crises became a topic of burning public interest. The felling rate throughout the country could no longer be allowed to in-

crease. A natural result of this was that greater effort was invested in the rapid and efficient renewal of forest resources. In the mid-1960s, the Company produced a maximum of six million seedlings a year. About ten years later, the figure had increased to 20 million. The organization in charge of wood purchasing was reorganized radically on the basis of cooperation between four forestry companies in central Sweden.

Finally, it should be noted that fears of over-felling were never realized, since the critical level was never attained.

Power

The expansion of the hydropower facilities on the eastern branch of the Dalälven River continued throughout the 1960s. Parallel with this development the Company had, since 1964, acquired substantial interests in a number of large heating plants which had been designed and constructed in other parts of the country.

However, few tempting opportunities remained to build additional hydropower plants on the Dalälven River system.This did not, however, mean an end to expansion of the Company's hydropower facilities. During the first year of the 1970s, a third unit was added to Trängslet, and the power station at Kvarnsveden was expanded. "Power," said Sundblad on one occasion, "is the source of the Company's strength".

The Steel Crisis

Towards the mid-1970s, it was time to raise the curtain on another change of scene. The year 1973 was marked by runaway inflation. Prices rose for many export goods, including forest industry products and steel. But import prices also increased. Against this background plans were laid for a new, large-scale state-owned steel mill in Luleå, in addition to that

already in operation in the far north of Sweden. In 1974 the authorities approved the construction of 'Stålverk 80' (Steelmill 80) which was to be designed with enough capacity to produce four million tons of billets a year.

As these plans were maturing, the international market was rocked to its foundations. In one fell swoop, OPEC had raised the price of oil four-fold. None the less, 1974 was a favorable year for Swedish industry. 'Excess profits' in the industrial and commercial sectors became a hot topic.

Wage demands increased drastically and met only weak resistance. Total labor costs increased almost overnight by 40 percent. Sweden suddenly found itself at the top of the international wages league without any corresponding increase in productivity.

At the same time, the international crisis deepened. In 1976 steel exports from the EEC countries had fallen by two-thirds in two years, while imports increased substantially.

The primary reason was not—it was believed—unfavorable economic conditions, but the fact that plant in the mature industrial countries was outdated and small-scale in relation to the modern steel mills of Japan and the newly industrialized countries. Furthermore, the mills in the mature industrial countries had higher wage costs and more costly raw materials.

During the period dating from the mid-1940s up to the beginning of the 1970s, Sweden managed to increase its output five-fold. The oil-price shock reversed the trend. When the international slump struck Sweden with its full force, the nation was poorly prepared. Sweden was more adversely affected by the crisis than many other countries. The failure of national economic policy was one reason. Another was the orientation of Swedish industry. The new situation on the oil market meant that, suddenly, the international tanker fleet was grossly over-dimensioned. The Swedish shipping industry, which was already struggling with major problems of its own, witnessed a further drastic deterioration in the situation. New orders for ships dried up. The numbing misfortune of the shipyards was soon to repeat itself in the steel industry. Painful reorganization was unavoidable.

In an address which he gave early in 1976, Bo Berggren, at that time deputy managing director of the Company, analyzed and at the same time provided a panoramic view of the steel situation and its relevance to Stora's future. This address focused on the fact that industry was already suffering substantial losses and that no solution was in sight. Continuing, Bo Bergren offered a number of proposals aimed at improvement, with special emphasis on marketing. Much of what was proposed on that occasion has since been implemented, but in retrospect the address, which aroused considerable interest at the time, is perhaps most interesting as proof that Stora's management had carefully analyzed the situation and were also prepared to take action.

The choice was relatively simple: make large, high-risk investments or simply divest the Company of its steel operations. There were many arguments in favor of the latter alternative, but it was not an easy road to travel.

The first problem area to be tackled was the specialty-steel mills at Söderfors and Vikmanshyttan. The management of the two mills had drawn up a reconstruction plan. It was highly optimistic. The mill would be able to achieve profitability without the need to merge with others. What was required was additional investment and an increase in operating capital which together would amount to some SEK 200 million. Considering the many years of losses which preceded this proposal, it is easy to sympathise with the doubts of Company management regarding such plans.

Other suggestions came from a survey in 1975 carried out by the Swedish Ironmasters Association. The objective had been to create a basis for a discussion about the possibility to achieve rationalization of operations by means of restructuring within the industry. An outline of the product programs of the various special steel companies showed that

Stora and Uddeholm had most in common and, thus, the most clearly competitive relationship. Combined, they would be the country's largest manufacturer of various types of tool steel. A merger would most likely lead to substantial profits for both companies.

It was therefore relatively easy to reach an agreement in principle concerning a merger between Uddeholm's Hagfors plant with Stora's two mills. Since Uddeholm was the larger manufacturer it was reasonable that it should acquire Stora's specialty-steel operations. The basic principles of the agreement, which was worked out at Stora, were that the mills in question should be transferred for a low price. This was a reasonable suggestion, since the Company had not been able for a long time to extract a reasonable return, even on restricted working capital. On the other hand, Stora would receive payment for its inventories and would retain its accounts receivable.

After three months of negotiations, agreement was reached as to how the plan should be implemented.

It was only at this point, in September 1976, that the plan was made public. Both the personnel and the municipalities concerned were taken by surprise. The many protests and proposals for alternative solutions were vigorously aired. Consequently, it took much longer than expected to reach a final decision.

The most awkward problem was the fate of Vikmanshyttan. This ironworks had had to be left out of the final agreement since Uddeholm had no reason to acquire it.

This ironworks was one of the smallest units of its type in the country, consequently operations were threatened even under normal conditions. As a result of the prevailing situation, it seemed likely that only a very small part of original manufacturig activities, the rolling of speed-steel strip, would remain. Other sections would have to be closed. At best, the personnel would have to count on finding employment within new trades.

It was at this stage that what is known as the 'Åsling doctrine', named after the then Secretary of Industry, was formulated. This stated that it was primarily the task of the owner, in this case Stora, to ensure that alternative employment was found for workers threatened by layoffs in a particular industrial area. An intensive effort was therefore launched, and bore fruit. Not counting the cold-rolling mill, which was to be operated by Fagersta, six new companies were established at the site. The prospects of rescuing Vikmanshyttan as an industrial community looked bright in 1977.

Company management quickly reached the correct conclusions from analysing a crisis which was still a long way from its peak. Being the first to react, it proved possible to implement a streamlining program which was far less drastic than those which eventually had to be implemented by companies which had hesitated to take action at this early stage. The price to be paid was a vicious public debate.

When other company closures followed at a later date, public reaction was muted; general opinion had become resigned.

"An ironworks is never finished", said Håkan Abenius in connection with the completion of a stage of the expansion of Domnarvet at the beginning of the 1950s. This was never more true than in the case of Domnarvet. Here, as noted earlier, reorganization and expansion projects continued uninterrupted.

Cooperation with The Gränges Company had generated many positive results, most importantly the elimination of the risk that its works at Oxelösund might enter the market for thin sheet, a market of vital importance to Domnarvet.

When the ambitious plans for Stålverk 80 were announced, the Company felt that the old threat had been resurrected in a more serious form. NJA (Norrbottens Järnverk) would soon realize—it was believed—that the large-scale manufacture of billets would have to be followed by further value-adding processes. Thin metal sheet, a profitable product, would offer a convenient alternative. Stora's solution to this dilemma was the proposal of a joint rolling mill at Gävle. After the Stålverk 80 project was finally shelved in November 1976, the rolling mill was also removed from the program.

But it was not just a matter of guarding the profitable sections of the ironworks. Efforts were also made to attain more direct contact with customers, and radical measures were adopted to divest unprofitable products. Stainless steel sheet production was sold to a German company, and heavy steel sheet operations to Gränges.

Investments designed to improve the blast furnaces at Domnarvet, which were made over a period of several years, were finally completed in December 1976. They were therefore in good condition, although still on the small side. However, the next stage of the production process, blast-steel production, was conducted in plants which could only be described as out-dated.

Other parts of Domnarvet, such as the scrap-based electro-steel mill and the equipment for the manufacture of various grades of thin metal sheet were technologically highly competitive and were therefore never questioned.

None the less, the ironworks was a great burden for the Company in the mid-1970s, with annual losses running into hundreds of millions of Swedish kronor, while the severe recession continued. At the beginning of 1977, Domnarvet manufactured for stock until all available space was filled. A large section of the workforce embarked on training programs. Radical measures were unavoidable, and they had already been drawn up. These were submitted as a series of concrete proposals included in the survey noted above, which was carried out by Lars Nabseth. The conclusion was that the three largest commercial steel mills in the country ought to be merged to form a single company. The capital investment required by such a solution was beyond the means of any private company. State involvement was unavoidable.

Stora's management had had close contact with Nabseth during the preparation of the survey and had contributed considerably to the substantial amount of information on which it was based. The recommended action program coincided with the guidelines drawn up by Stora. That they were well founded is confirmed by the fact that they were generally accepted as the basis for the formation of the new joint steel company.

Roles within the new company were to be distributed in the following manner. Oxelösund would be the sole manufacturer of heavy plate, which was already a fact. Domnarvet would be the only producer of hot-rolled strip and cold-rolled sheet. As had previously been the case, some of the sheet would be galvanized and coated. NJA would manufacture bar steel in large dimensions.

In general, the prospects for sheet products were regarded as being better than those for bar.

This applied in particular to cold-rolled and galvanized sheet. The reasons for this were the growing market, the great potential for further processing and the substantial import share which existed at the time of the survey. Consequently, despite its losses, Domnarvet was "the best placed of all Swedish steel mills" (Sundblad).

In the case of ore-based metallurgy, it was stated that Domnarvet had unsatisfactory equipment, while NJA and Oxelösund were both suffering from over-capacity. In this respect, Domnarvet found itself in the danger zone.

In terms of personnel, the proposal meant that Domnarvet would face the largest reductions, with a total loss of about 2,000 jobs. In addition, there were the employees at the company's mines, totalling about 500 people.

But the plans were not entirely negative. The survey also proposed an expansion of production capacity at Domnarvet. This applied in particular to hot-rolled strip. (Under the direction of the new Company, this gradually developed into a multi-billion kronor investment.)

When the survey was published, the management of Stora supported its basic proposals without reservation. Other parties involved were also positive, but this did not prevent negotiations from being protracted and complicated.

Conditions within the three companies differed considerably. As far as Stora was concerned, the steel operations were a great burden, but the Company was not hopelessly tied to the business in its present form. There were alternatives. Furthermore, major parts of the Company were untouched by the crisis and had excellent prospects for survival.

Gränges declared that it was completely crippled. The company had sustained catastrophic losses in 1975 and 1976. Even before the commercial steel survey was complete, Gränges made it clear that it would welcome a state takeover of Oxelösund.

Finally, NJA, controlled by Statsföretag (the Swedish State Holding Company), had hardly ever shown a profit but could nevertheless count on state assistance, mainly because of the social implications.

After protracted, complicated and sometimes heated negotiations, a final agreement was signed at year-end 1977.

The state contributions were justified on the basis that it was necessary for the country to have its own steel manufacturing operations, especially with a view to supplying the engineering industry, and in view of the fact that as many jobs as possible had to be saved. It was also evident that the steel industry could not survive in its current form, since this would have made it impossible to generate the resources required for essential investment.

An important question for Stora was the question of what payment regulations would apply. It seemed not unlikely that the new company, for social and other reasons, would feel it necessary to deviate from established principles of profitability. If the private owners could not influence such a development, they should then have the option to withdraw. A provision to this effect was accepted by the parties involved. These fears were confirmed only too soon. Three years later Stora had every reason to sell its shareholding. The price received was one Swedish krona.

At the beginning of 1978 Stora could feel satisfied with the result of the negotiations. The Company had managed to divest itself of a business which had incurred substantial losses without entailing unacceptable financial or social consequences. The Company's profitable sections and those parts capable of development remained intact.

Stora had adopted a new role. In an address delivered in 1978, Sundblad said: "Following the restructuring operation we have undergone, there is no doubt that Stora Kopparberg has become one of Sweden's strongest forest-products companies. It is now up to us to ensure that we create something really good, profitable and expansive from the material we hold in our hands."

Even though the major problems in the steel sector had undoubtedly acted as a serious obstacle to the Company's development as a whole, the recovery was not long in getting under way. After only two years, Stora was able to report earnings which ranked it the "best among the country's forest-products industries". The profit reported was the largest in the history of the Company.

The road ahead was now clear, offering the freedom to strive for new objectives. The next ten years were to see the development of a completely new company.

STORA

THE MEN WHO PLANNED Kvarnsveden Paper Mill ninety years ago made a number of good decisions. One of these was their selection of its location and layout. The decisions taken then made it feasible to add new, large units, one after the other, without the need for any major conversion of the original mill facilities.

The most recent addition, as we enter 1988, is the new newsprint machine, Number 11, with an annual capacity in excess of 200,000 tons. Its completion coincides with that of one stage in the expansion of the Grycksbo mill. In this case, the product will be coated fine papers. These are not isolated examples. Similar investments are in progress or being planned at half a dozen other facilities in Sweden, Portugal and North America. But this is not merely a question of large plants becoming larger.

In just a few years, the Company has acquired an entirely new profile. The most comprehensive changes were implemented around the mid-1980s.

In 1984, the Company acted with what outsiders considered 'lightning' speed when it acquired Billerud, the flagship of the Värmland forest companies. Parallel with this transaction, Group operations in North America were expanded with the acquisition of the Newton Falls paper mill in New York state. The mill was to produce fine paper based on the same program as that introduced at Grycksbo with such notable success. It was also to become a major purchaser of sulfite pulp from the Group's plant in Nova Scotia.

Two years later, STORA expanded further, following a merger with the large west-Swedish forest-industry group headed by Papyrus. This was the largest corporate merger in Sweden up to that time.

In the same year (1986), Stora was able to realize a considerable sum following the signing of a partnership-financing agreement with a consortium of Swedish insurance companies and pension funds, which involved transferring control of the majority of the Group's hydropower assets. This transaction was made conditional on the understanding that STORA maintain a majority of the voting rights in the new company, Kopparkraft, which was formed. Billerud's power assets were sold at the same time. These measures led to a considerable improvement in the Company's financial position.

Stora has now become the largest forest industry company in Sweden, and one of the ten largest in the world.

Combined, the newly acquired units represent the largest part of the Group as a whole.

This major expansion of Group activities was not the result of some sudden enthusiasm in the idea of growth as such. Rather, it has been the almost inevitable result of international economic and technical developments.

The market's growth and the new grades being demanded by customers have necessitated new investment by producers. Larger units not only yield a better return, they also normally yield better and more uniform quality. However, they also require greater capital. The sort of capital required could only be generated by large companies. An expanding product range and a larger number of manufacturing units provide a more comprehensive pool of experience and substantial synergetic benefits. All aspects of manufacturing ought to expect to gain from increased resources. This applies as much to R&D and production as to purchasing, transport and marketing. This also increases the Company's ability to meet various environmental demands.

The "new Stora" can market products with a higher added value than has previously been possible. There are excellent opportunities for reducing the time taken to convert a raw material into a paid-for, finished product. Profitability can be improved while reducing sensitivity to fluctuations in the business cycle.

The new companies

The comprehensive nature of these acquisitions has also had a certain spinoff effect: STORA has suddenly become the heir to new and significant portions of Swedish industrial history.

Through the acquisition of Billerud, Stora became a participant in the varied patchwork of Värmland's industrial traditions.

Billerud traces its origins from 1883. This was the year in which Viktor Folin started to build a small pulp mill outside Säffle. Folin was an eminent cellulose technologist. He had learned how to produce sulfite from one of the inventors of the process, Alexander Mitscherlich. Folin was for the time being the only one in the country to succeed in making his plant profitable relatively soon after its startup. This was a pioneer venture: Billerud was the third sulfite plant to start up in Sweden.

Like Superintendent G.T. Lindstedt, Billerud's first administrative director, Folin came from Munkedal. K.A. Wallenberg, the banker, provided vital support.

By 1888, the plant was producing 2,000 tons a year, which gave the company a leading position among the country's chemical pulp producers.

It was at this early stage that Folin, having accepted an assignment from K.A. Wallenberg, left Billerud to build and then successfully manage a sulfite plant at Storvik, in Gästrike-Hammarby.

However, Billerud was to become a really large company under the leadership of the well known Christian Storjohann. He was appointed president of the company in 1907, and led it for 40 years before relinquishing his position to become chairman of the board. During his long presidency, the company acquired considerable forest assets and many production facilities, comprising 20 companies. Especially noteworthy years were 1922, at which time the company started what was to be periodically an extremely successful production of tissue pulp, and 1932, when it started to promote a new grade—bleached kraft paper—used in packaging. Four years later, the company began to produce paper sacks. From the beginning of the 1930s, the sulfate plant and the paper mill at Gruvön were to prove crucial links in the company's chain of production facilities. In parallel with the closure of operations elsewhere, Gruvön was developing into one of the largest forest-industry plants in Europe.

Billerud's decision to establish a company in Portugal—Cellulose Billerud SARL (Celbi)—in 1964, was to prove significant for the company's future. A plant was started up in 1967. The raw material was eucalyptus.

The plant developed into the foremost of its kind in Europe. As a result of this success, Billerud was commissioned to supervise construction of the world's largest eucalyptus-based pulp mill, in Aracruz, Brazil. Billerud also became a minority shareholder in the company. The mill started production in 1978. Around 1980, Billerud had developed into a world leader with respect to fluting (the corrugated layer in corrugated board), liquid packaging and sack paper. Corrugated board was produced at ten board mills on the Continent. At this time, the company had sales of approximately SEK 4 billion, and 8,000 employees.

Uddeholm's forest-industry operations, which had been merged with Billerud in 1978, accounted for a significant proportion of the company's activities at this time.

Uddeholm had for a long time been the Swedish company most like the old Stora Kopparberg in terms of orientation and organization. Like the Stora Kopparberg of the 1800s, Uddeholm's business activities developed from a base in mining, iron works, forests and forest industries.

It all started in 1668, with the construction of Uddeholm's first iron works, comprising a blast furnace and forge. The company expanded. A number of iron works were acquired or newly constructed over a period stretching from the 1820s to 1939. At this time, five large units were in operation, all of them close to, or east of, the Klarälv River in Värmland.

In the mid-1950s, the company sold steel products to a value of SEK 200 million. This was approximately the same amount as that reported by Stora Kopparberg. At the same time, Uddeholm's invoiced sales for wood products, chemical pulp, paper and board amounted to SEK 157 million, while Stora Kopparberg's sales for the same products exceeded this amount by about SEK 10 million.

Uddeholm's major forest complex was located in Skoghall at Hammarön, south of Karlstad. The history of this complex dates back to the beginning of the 1830s, when an Iron Scale had been erected at Skoghall, to be used when unloading the works' products. A small sawmill was also located here, dating from 1855. A radical concentration of the company's forest industries to Skoghall was implemented during and after the First World War. A new saw was constructed in 1914, a sulfite plant in 1917, a sulfate plant in 1919 and a paper mill in 1930. To these were added an electrochemical plant and a board mill. The result was a complex able to process the full range of forest products, everything from large logs to the simplest knot pulp. In the 1950s, Skoghall was the largest forest-products processing complex in Europe.

Uddeholm received a boost in 1967, when the company acquired the Mölnbacka group of companies, comprising forest assets, power stations, sawmills and pulp- and paper mills, from the Norwegian company, Borregaard. AB Mölnbacka-Trysil had been established in 1873, but its period of real growth did not occur until towards the end of the century.

As for Skoghall, it had got into serious difficulties at the time of the merger with Billerud. A large, newly built boxboard machine proved to be afflicted with a number of weaknesses and faults. Following the acquisition, it was converted by Billerud for the production of liquid-packaging board.

Uddeholm's steel manufacturing operations were radically restructured in cooperation with other Swedish companies in the steel business.

The Stora Kopparberg-Billerud integration, subsequently always referred to as STORA, involved substantial synergetic benefits. This integration was to prove even more natural and beneficial when the old dream of a merger between STORA and Papyrus was realized in 1986.

Ever since Papyrus was founded in 1895 on what was left of Korndal, a company which had gone bankrupt, it had held a special status within the Wallenberg group of companies. This was Deputy District Judge Marcus Wallenberg's first major industrial initiative, an initiative he took in his capacity as legal counsel to Enskilda Banken (a major Swedish bank working closely with several large Swedish companies). He relied on Victor Folin as a consultant.

From this time, Papyrus built its business from the ground up. The company produced fine paper and board, as well as limited quantities of sulfite and mechanical pulp. Due to skillful mangement, the company was soon profitable and production increased.

Industrial traditions in the town of Mölndal dated back to 1653, when the first paper mill was established. Korndal could trace its ancestry back to the latter part of the eighteenth century.

The First World War and the years which followed in its wake were years of crisis for Papyrus as well as for the rest of the industry. When a new machine for the production of boxboard was started up in 1927, it represented the first major investment for a very long time.

A period of expansion began during the latter half of the 1940s. It was at this time that the company acquired Yngeredsfors, a power company, which owned forest industries in Hylte and Oskarström on the Nissan River, to the south of Mölndal and Gothenburg. This involvement in power production fluctuated over the years, and was not finally discontinued until the beginning of the 1980s.

The Mölndal mill was twice extended, which enabled it to establish a commanding position as a manufacturer of coated packaging board. In the 1960s, the fine-paper mill was expanded.

Dating from 1967, close cooperation had been established on the sales side between Grycksbo, Munkedal and Papyrus. A few years later, the various wholesaling companies owned by these three were merged to form a single, joint-owned company: Pegenova. This company now handles sales of fine papers for the STORA Group as a whole.

Two other plants, at Nymölla and Hylte, also constituted important aspects of Papyrus' development.

Nymölla AB was formed in 1960 on the initiative of Marcus Wallenberg (the son of Marcus Wallenberg, Senior). The company was established primarily to meet Papyrus' requirement for a pulp supplier. Southeastern Sweden possessed large, although as yet mostly unutilized forest assets, consisting primarily of hardwood and some conifer. With this in mind, the decision was taken to invest in a new method, which could be applied to every type of wood available in the vicinity, to be introduced at a new unit which was being planned just east of Kristianstad. The method, at this stage unproven on a commercial scale, involved the use of magnesium bisulfite in the digesting liquor.

The plant started operations in 1962. It was to cause its owners considerable problems. The plant lost money year after year, virtually throughout the 1960s. It was not until 1969 that Nymölla attained a degree of financial, technical and managerial stability. By then, the company had achieved an annual production of 110,000 tons of 'magnefite pulp'. At that time, the method had also achieved international recognition. The following year, the company felt itself to be in a sufficiently strong position to double capacity at the pulp mill. The majority of Papyrus' chemical pulp requirements could now be supplied by Nymölla, where a paper mill was also constructed.

The trend has remained favorable. A program of expansion comparable in size to that carried out at Kvarnsveden is currently being completed. The end product will be coated fine-paper.

Nymölla was originally formed as a subsidiary to Hylte Bruk, in which Papyrus had a 60-percent interest at the time of the construction of the new plant.

Hylte itself achieved little success during the 1960s. Like Nymölla, it suffered many losses. The plant was in a poor state of repair, and completely unsatisfactory with respect to environmental considerations. However, as already mentioned, a turning point was reached in 1970, primarily in response to an initiative from Erik Sundblad. As a result, a new Hylte Bruks AB was created, with Marcus Wallenberg as the driving force, and with considerably greater resources. Hylte was to install what was Europe's largest newsprint machine at the time, with an annual capacity of 165,000 tons. To this would be added a mechanical pulp mill and a plant for the production of unbleached magnefite pulp. The consortium which provided the financial backing for this project aroused considerable interest: apart from the original owners, who supplied slightly less than half the total capital required, the consortium included a major German paper producer, Feldmühle (25 percent), as well as the state-owned forest-industry company ASSI (20 percent), which later sold its interest—and Stora Kopparberg.

The new company was a success. Since then, the company has stuck to its policy—to produce newsprint, and nothing but!—with dogged persistence. During the 1970s, Hylte Bruk installed two more large paper machines. Another paper machine is under construction. When this comes on line, Hylte will probably become one of the world's largest newsprint production complexes located in one place. Following the implementation of a consistent policy of environmental safety, the Nissan River downstream of the mill now offers excellent angling. The mill's substantial and growing use of waste paper provides another positive environmental contribution.

In the same year that Bergvik and Ala were merged with Stora Kopparberg, 1976, Papyrus acquired the shares in Kopparfors, a company which, at the time, in addition to almost 200,000 hectares of forest, also owned a sulfate pulp mill and a sawmill in Norrsundet on the coast of Gästrikland, a plant for the production of unbleached magnefite pulp at Gästrike-Hammarby, and a board mill in the old industrial town of Fors, in the southernmost part of Dalarna.

Kopparfors AB (known as Kopparbergs & Hofors Sågverks AB until 1937) was formed in 1855 as the result of an agreement between the owner of the Hofors works, Stora Kopparberg, and a limited number of individual investors. As mentioned earlier, Stora's relationship with this new company had been close, especially during the first few decades. In 1887, the company had acquired what was in certain respects a new profile, as a result of the acquisition of what was then known as the 'Ockelbo Works'. In addition to increased forest assets, this resulted in fairly substantial involvement in the iron processing industry, and a transfer of the geographical center of operations further to the north and to the coast.

During the last few years of the nineteenth century, members of the Nisser family were the sole owners of the company. This picture changed radically when Hofors AB, a member of the Wallenberg group of companies, entered the arena as a 50-percent joint owner of the company in 1906. A few years later, Kopparfors found itself facing a serious crisis brought on by incompetent management. In 1909, the company was forced to report a substantial loss, while at the same time burdened by heavy debts. It was saved from ruin by Emil Lundqvist, who joined the management at this point and rapidly managed to reverse the downward trend. The business started to make a profit and its finances were restructured.

When Lundqvist left the company at the beginning of the 1920s to head Stora Kopparberg, Kopparfors had the resources to expand. Both the facilities at Hammarby and the newly established sulfate plant at Norrsundet were subsequently expanded. Iron operations, however, were discontinued.

In 1981, Feldmühle acquired a 50-percent share in a newly formed company, Norrsundets Bruk. At Fors, which was retained within the company, the plant concentrated on the production of chemi-mechanical pulp (CTMP) and expanded its boxboard manufacturing activities. A new plant for plastic coating of boxboard from Fors and Mölndal was built at Hammarby.

The new Group

The job of bringing together the motley collection of companies and facilities described here, with their shifting relationships and different traditions, and of turning them into a single new and efficient unit was implemented more logically and consistently than at first sight might appear possible. In many cases, as described above, this process had been preceded by a tradition of personal contact and many years of cooperation.

In principle, Group activities were organized according to product orientation. This approach was applied to the two large raw material divisions, forest and power, whose management was located at head office in Falun. This also applied to sawmill- and joinery operations, which were directed from Söderhamn; chemical pulp production (headquartered in Gävle); fine paper operations, directed from Papyrus' Mölndal office, and the division responsible for the manufacture of products such as packaging paper and boxboard, concentrated mainly at Billerud. The large newsprint mills at Kvarnsveden and the one at Hylte, in which the Group now held a 75-percent interest, were organized into independent units, as were pulp- and paper operations in Nova Scotia, Canada. To these were added five smaller units.

Group forest assets now amount to 1.6 million hectares, corresponding to approximately one third of the raw material requirement. Raw material purchasing has been simplified by entrusting this activity to a limited number of large wood-purchasing agents in different parts of the country.

All the Group's primary products may be said to offer good prospects for growth. Fine paper is considered a priority growth sector, although liquid packaging, as well as newsprint and magazine paper, eucalyptus pulp and non-woven fabrics are also judged to be growth products.

Chemical-pulp production is supplied increasingly to Group manufacturing units where value is added, although STORA remains one of the largest actors on the market, producing about a million tons of market pulp a year. One of the most successful grades is the fluff pulp used for hygiene products. Capacity has recently been expanded, both at the Skutskär and Skoghall mills, to meet demand in this sector.

The trend for newsprint has been favorable. Dating from 1982, the world's newspapers have become increasingly comprehensive, with ever-thicker weekend and special supplements. The number of advertisers has grown. Group production facilities are able to meet the market's increasingly stringent demands for paper grades which can be used in a variety of printing presses and for different printing processes, as well as demands for reduced surface weights.

Various types of packaging paper also represent a key item in the balance sheet. The STORA Group produces both the corrugated inner layer and the visible outer layers used in the manufacture of corrugated board. Sack paper is a product which faces stiff competition from other packaging materials. The market for standard grade sawn-timber products is weak: orientation towards a product range with higher added value is required.

Several of STORA's smaller units are able to report strong demand for their products. One example is G-man, a producer of printing inks. Another is Storalene, a manufacturer of non-woven fabrics.

In this varied range of industries, which nevertheless operate in related sectors, the division which now bears the company's old name, Stora

Kopparberget (The Great Copper Mountain), plays the role of odd man out. It is also a small division. Despite this, iron pyrite concentrates still provide the basis for a profitable chemical industry, and 'Falun Red' paint still sells to a loyal group of customers.

This is not all. The Falun Mine, even after being worked for a thousand years, can still provide some welcome surprises. It may well be that gold ore could once again come to play an important role in making the mine profitable, as it did a hundred years ago. However, it is important to remember that extraction of the remains of the original deposit is a complex process.

Doubts

Elements of risk are to a large extent shared by other Swedish forest-industry companies. At least five problem areas are easily defined: the competition from electronic media, the limited raw-material base, Sweden's indecisive energy policy, the trend of domestic costs and doubts surrounding possible future cooperation with the rest of Western Europe.

Electronics have affected paper consumption in a way that was not expected. Ever since the introduction of the telegraph around 1840, the capacity of electronic transmissions has consistently been doubled at intervals of less than four years. Never before in the history of the world has new technology made such a rapid breakthrough. It is obvious and inevitable that the new media will come to play an increasingly dominant role. Paper is being replaced in many contexts by diskettes. But in everyday operations, paper is still the main medium, as it is when it comes to long-term storage of information in archives. Nevertheless, even a shrinking share of society's total information requirement will still mean not inconsiderable growth in overall demand for paper. It seems fair to assume that the amount of paper used will continue to increase in pace with the rise in living standards.

In addition, certain types of information are definitely better on paper than on computer screens. These include advertisements and various forms of practical manual (such as user instructions for electronic equipment). Essays and contributions such as those found in the leaders, arts and literary columns of newspapers, as well as memoranda and other 'unstructured' material, are also best consigned to paper. You can seldom make notes in the margin of a computer screen. And it would be even more difficult to take it to bed with you, to ponder on it a bit more.

To put it another way: the more important the message, the more important that it be consigned to paper.

It is also worth noting that the 'computerized office' appears rather to have increased the amount of paper generated than otherwise. Moreover, this applies particularly to the special grades, which command a high price.

Computer technology has made the production of paper-borne products cheaper and faster. This also means that these products are likely to be in greater demand than before. It is now possible, for instance, to prepare printed matter or a book in the consumer's home, rather than at the printers. The product can be transported via a fiber-optic cable, instead of by truck. Paper used in packaging can generally look forward to a growth market.

There is every reason to predict continuing demand for Swedish forest products in the long-term.

The problems are of another kind. They may arise as the result of over-capacity, or due to fluctuations in the business cycle, which experience has shown can have a particularly negative effect on the forest industry. The problems faced by the industry may also be of a more basic nature, such as the supply of raw material.

Sweden—and more specifically STORA—possesses forests which are perhaps better suited as raw material for the production of highly-refined paper grades than those found anywhere else. There is actually more Swedish timber available today than earlier. Thanks to conscious efforts, the nation's forest resources have doubled over the past hundred years, fairly surprising in a world characterized by the ruthless exploitation of forests, and by their death from pollutants.

Since the 1960s, Sweden's forest industry has achieved a dramatic increase in productivity, but the Swedish raw material is still more expensive than that from other countries, and availability will continue to be limited. In consequence, Sweden has become a pioneer in developing methods for producing wood pulp which optimize the raw material available. The new thermo-mechanical processes utilize almost all the wood entering the process. The paper is suitable for a broad range of applications. These processes also offer a high level of environmental safety. This is why newsprint, for instance, is produced to an increasing extent from thermo-mechanical pulp and insignificant quantities of sulfa-te. The new high-yield pulps can also be used to advantage in the manufacture of products such as liquid-packaging board and fluff pulp.

This is without doubt a major step forward. There is, however, a complication, arising from the fact that these new processes require large amounts of energy. This fact serves to emphasize a basic prerequisite for the continued existence of Sweden's forest industries, and an opportunity to further improve its products: access to large quantities of electrical power at a reasonable price. In this respect, the future remains uncertain.

The industry's own solution to the problem of limiting the increase in the use of electricity has been to improve efficiency as far as possible, and to integrate production. Sharpened demands on product quality, and care of the environment, have nevertheless led to steady growth in the demand for energy.

Environmental questions

For a long time, the West has endorsed the idea, formulated in the first chapter of the Old Testament, that the Earth was created for Man. This concept was formulated with particular acuity, at the beginning of the seventeenth century, by Francis Bacon. Mankind ought to try to dominate Nature. Nature was to be formed in compliance with Man's wishes, and thus persuaded to give of its best.

This attitude has been enthusiastically endorsed by subsequent generations. Mankind has worked hard and assiduously at wearing down the Earth on which he lives.

But Bacon also emphasized that the best way to control Nature was to respect its laws. It is this realization which is now becoming more generally recognized. Technology has created much of our wellbeing, but technology which destroys the environment is meeting increasing resistance. Care of the natural environment should not, however, imply a ban on technological development. Perhaps the *really* great technological breakthroughs are still to be made, such as those which enable superconductors to be used at more normal temperatures, or which would allow us to generate hydrogen power or store solar energy.

Due to the restrictions imposed by Nature herself, industries which rely on the forests are forced to adopt longer-term planning than other commercial sectors. Care of the land, the water and the air is a question of self-interest. No other Swedish industry has invested anything like as heavily in environmental care as the pulp and paper industry. Reductions in the atmospheric emission of sulfur dioxide and the release of oxygen-absorbing substances into water provide examples of two areas where dramatic advances have been made. In both cases, reductions in the range 80 to 90 percent have been achieved since the 1960s. In the few remaining years of the 1980s, the sulfate plants alone will invest more than SEK 2 million in improved protection of the environment.

The greatest problem has been to deal with chlorine-rich effluents from bleaching plants. The Gruvön and Skutskär mills were among the first to introduce oxygen bleaching, a great step forward. This is now standard throughout the country. The use of chlorine as a bleaching agent has been halved, and will certainly be reduced even further. In this field, Sweden has a significant lead over other producing countries. STORA's basic policy with respect to long-term environmental questions is clear: operations are to be carried out in a manner which will have no adverse effect on the environment, and which will pose no threat to future generations.

However, such measures are not uncommonly extremely costly. It is therefore important to ensure that new methods are thoroughly tested before they are adopted on a commercial scale.

The benefits of history

Pondering over the way of the world and trying to sift some practically useful insights from the river of history was a favorite pastime among the learned men of earlier times. Tales about the achievements of great men of the past could be used to encourage people to all sorts of virtuous deeds. In his chronicle, Olaus Petri, whose contribution has already been mentioned, wished to give his readers the opportunity to "behold the way of the world, its vanity and instability, and the wiles of the Devil". For Olaus, the battle between good and evil was the main theme of history.

Whether such beliefs are justified is a matter for debate. One thing is certain, however: history cannot be used like a cook book. It cannot provide any reliable recipes for success. Nor can its content be rigidly interpreted. The history of the leaders differs from that of the led. Reality is largely in the eye of the beholder.

Seen in this perspective, history really has very little to contribute, apart from broadening one's insight into the relative nature of 'great truths', the multiplicity of options, the difficulties of distinguishing the undercurrents in what appears to be a consistent chain of events. It may serve to give a clearer understanding of the origin of the many puzzling and irrational forms of behavior which characterize today's society. Knowledge of this sort may perhaps help to make one feel more at home in a changing world. At best, history then becomes what the learned Lord

Acton of Aldersham wished for: "not a burden to the memory, but a vision for the soul".

In contrast, an attempt to provide concrete predictions about future developments based on a knowledge of history is a lot more difficult. A major study of anticipated developments made by a research team from the American National Resources Committee, published in 1937, failed to predict scientific breakthroughs such as television, atomic power, radar technology, the jet engine and the computer, all of which had become a reality ten years later.

Methods of prediction may have been refined since then. Nevertheless, there is still reason to suspect that it will be the unexpected that will occur. At the same time, it seems reasonable to assume that most of what is likely to be everyday technology in the year 2000 should be discernible today from trends and intimations.

Internationalization moves rapidly and implacably forward. How this will affect Sweden remains in some respects an open question.

The Swedish forest industry, and STORA especially, are well prepared to make their contribution to further this development. They are linked to the surrounding world in thousands of different ways.

*

When work on this book was nearing completion, STORA initiated yet another major business transaction, which will have far-reaching consequences on the entire structure of the Group.

The proposed acquisition of Swedish Match would lead to substantial integration: at the same time, STORA would enter a number of new areas of operation. The Group's presence on international markets would also be consolidated and extended.

STORA can look forward to its eighth century with substantially greater breadth and strength than previously.

Postscript

While researching the material for this book, I have received invaluable assistance and good advice from a number of experts within STORA. In particular, I should like to mention Archivist Rune Ferling, who has sacrificed many hours to reading and commenting on my manuscript.

Outside the Company, I have had the good fortune to be able to discuss the entire work with Sten G. Lindberg, Ph. D.

Mrs. Marianne Fornander has been a great help in allowing me to read her translation (including all notes) of Charles Ogier's recently rediscovered second account of his travels.

The Swedish Pulp and Paper Association has contributed some valuable insights, in particular by giving me access to the manuscripts of a number of lectures given by the president of the Association, Bo Wergens.

Bibliography

A complete and systematic account of the fairly comprehensive list of works which have provided the basis for this book seems hardly appropriate. A few key works should, however, be mentioned.

Key references

BERTIL BOËTHIUS: *Gruvornas, hyttornas och hamrarnas folk* (1951). – *Kopparbergslagen fram till 1570-talets genombrott* (1965).

ELI HECKSCHER: *Sveriges ekonomiska historia I – II,* (1935–1949). – *Svenskt arbete och liv* (1941) (also a later edition). – *Den svenska kopparhanteringen under 1700-talet* (Scandia 1940).

K-G. HILDEBRAND, *Falu stads historia I–II* (1946). – *Fagerstabrukens historia I, sexton- och sjuttonhundratalen* (1957). – *Erik Johan Ljungberg och Stora Kopparberg* (1970).

S. LINDROTH, *Gruvbrytning och kopparhantering vid Stora Kopparberget I–II (1955)*. – *Vetenskapsakademiens historia I–II* (1967). – *Svensk lärdomshistoria I–IV* (1975–1981).

The Middle Ages

NILS BJÖRKENSTAM, *Den gamla svenska masugnen* (Polhem 1985/3), as well as essays in *Medieval Iron and Society* (Jernkontorets bergshistoriska skriftserie H 34, 1985) and the article on Osmund in the Medeltidens ABC (1985).

T. K. DERRY & T. I. WILLIAMS, *A short History of Technology* (1960).

C. GREGORY, *A concise History of Mining* (1980).

KJELL KUMLIEN, *Svenskarna vid utländska universitet under medeltiden* (Historiska studier tillägnade Sven Tunberg 1942). – *Sverige och hanseaterna* (1953). – (ed.) *Norberg genom 600 år* (1958).

J. NORDSTRÖM, *Bidrag rörande Boetius de Dacia* (Samlaren 1927). – *Medeltid och renässans* (Norstedts världshistoria VI, 1929).

U. QUARFORT, *Sedimenten i sjön Tisken och Falu gruvas ålder* (Jernkontorets forskning H 19, 1980).

T. SCHMID, *Medicinsk lärdom i det medeltida Sverige* (Fornvännen 1951).

I. SERNING, *Prehistoric Iron Production* (Iron and Man in Prehistoric Sweden 1979).

The sixteenth century

I. ANDERSSON, *Svenskt och europeiskt 1500-tal* (1943).

S. J. BOËTHIUS, *Bönder, bergsmän och brukspatroner i Dalarnas historia* (Historisk Tidskrift 1916).

P.NORBERG, *Sala gruvas historia* (1978).

CH. SINGER (ed.), *A History of Technology III. From the Renaissance to the Industrial Revolution* (1957).

The seventeenth and eighteenth centuries

J. ALM, *Blanka vapen och skyddsvapen* (1975). – *Eldhandvapen* (1976). – *Vapnens historia* (1982).

G. HALLDIN (ed.), *Svenskt skeppsbyggeri* (1963).

L. HAMMARSKIÖLD, *Kopparkanoner i Sverige och deras utveckling* (Med Hammare och fackla XVIII, 1949/1950).

S. G. LINDBERG, *Från skapelsetro till naturlagar* (1959).

S. LJUNG, *Skultuna bruks historia 1607–1860 II:1* (1957).

E. LYBERG, *Borns hyttegård samt styckegjutarna i Falun* (1944).

L. MAGALOTTI, *Sverige under år 1674* (1912).

CH. OGIER, *Dagbok under ambassaden i Sverige 1634–1635*. Published by Sigurd Hallberg entitled Från Sveriges storhetstid (1914). – Ogier's second report of his visit published by Cristina Wis, *Una relazione del seicento sulle miniere del settentrione* (Studi Nederlandesi – Studi Nordici XXV, 1982, Istituto universitario Orientale, Napoli).

H. OLSSON, *Kemiens historia i Sverige intill år 1800* (1971).

G. ROSANDER, *Allmogens utnyttjande av skogen* (Dalarnas Hembygdsbok 1977).

Svenskt och utländskt järn på 1600-talets Europamarknad (Med Hammare och Fackla XXVIII, 1981).
Essays by i.a. MARTIN FRITZ and NILS BJÖRKENSTAM.

B. TINGSTRÖM, *Sveriges plåtmynt 1644–1776* (1984). – English version (1986).

The nineteenth century

A. ATTMAN, *Svenskt järn och stål 1800–1914* (1986).

G. ERIKSSON, *Kartläggarna. Naturvetenskapernas tillväxt och tillämpningar i det industriella genombrottets Sverige 1870–1914* (1978).

A. HASSELGREN, *Utställningen i Stockholm 1897*. Beskrifning i ord och bild (1897).

A. HELLDÉN, *Maskinerna och lyckan* (1986).

E. J. LJUNGBERG, *Om bolagsvälde och bolagshat* (1908) – *Vår modernäring och vår folkundervisning* (1914).

L. LOOSTRÖM (ed.), *Allmänna konst- och industriutställningen*. Officiell berättelse (1899).

G. SUNDBÄRG (ed.), *La Suède, son peuple et son industrie. Exposé historique et statistique* (1900).

R. ÅKERMAN, *Erik Johan Ljungberg* (Bihang till Jernkontorets Annaler 1915, häfte 6).

The twentieth century

I. ANDERSSON, *Uddeholms historia* (1960).

J. CALLERSTRÖM and G. NYLANDER, (ed.), *Företag i utveckling. Tillägnad Marcus Wallenberg* (1969).

E. DAHMÉN, and B. CARLSSON, *Den industriella utvecklingen efter andra världskriget* (Sveriges Industri 1985).

O. GASSLANDER, *Bank och industriellt genombrott. Stockholms Enskilda Bank kring sekelskiftet 1900 I–II* (1956–1959).

S. HÖÖK, (ed.), *Billerud 1883–1983* (1983).

L. NABSETH (ed.), *Handelsstålsindustrin inför 1980-talet* (SOU 1977/15–16).

U. OLSSON, *Bank, familj och företagande. Stockholms Enskilda Bank 1946–1971* (1986).

J. RENNEL (ed.), *Future of Paper in the Telematic World* (1984).